PRACTICE WORKBOOK

On My Own

Harcourt Brace & Company

Orlando • Atlanta • Austin • Boston • San Francisco • Chicago • Dallas • New York • Toronto • London

http://www.hbschool.com

Printed in the United States of America

ISBN 0-15-307932-0

10 11 12 13 14 15 16 082 2004 2003 2002 2001

CONTENTS

CHAPTER 9 Analyzing and Graphing Data

CHAPTER 10 Circle, Bar, and Line Graphs

CHAPTER 11 Probability

CHAPTER 12 Multiplying Decimals

CHAPTER 13 Dividing Decimals

CHAPTER 14 Measurement: Metric Units

CHAPTER 15 Understanding Fractions

CHAPTER 16 Fractions and Number Theory

CHAPTER 17 Modeling Addition of Fractions

CHAPTER 18 Modeling Subtraction of Fractions

CHAPTER 19 Adding and Subtracting Fractions

CHAPTER 20 Adding and Subtracting Mixed Numbers

CHAPTER 21 Measurement: Customary Units

CHAPTER 22 Multiplying Fractions

CHAPTER 23 Plane Figures and Polygons

CHAPTER 24 Transformations, Congruence, and Symmetry

CHAPTER 25 Circles

CHAPTER 26 Solid Figures

CHAPTER 27 Fractions as Ratios

CHAPTER 28 Percent

Name ________________________________

LESSON 1.1

Using Numbers

Vocabulary

Fill in the blanks.

1. ______________ numbers tell how many.

2. ______________ numbers tell position or order.

3. ______________ numbers name things.

Tell whether each number is expressed as a *cardinal, ordinal,* or *nominal.*

4. The area of Cheryl's room is 130 sq ft.

____

5. Laura won first place in the dance contest.

____

6. Frank scored 23 points in a basketball game.

____

7. Michael Jordan wears the number 23 on his jersey.

____

8. Mr. Lenny put a teaspoon of sugar in his coffee.

____

9. Amanda's mother is 39 years old.

____

Mixed Applications

For Problems 10–12, use the table.

WALK FOR HEALTH		
Winners	**Name**	**Miles Walked**
1st Place	Juan-Carlos	37
2nd Place	Sarah	34
3rd Place	Frank	32

10. What are the total miles walked by the 3 winners?

____

11. Frank walked at the same speed for 8 hours. How far did he walk in 1 hour?

____

12. If each mile walked raised $2.25, did Juan-Carlos earn more than or less than $75.00?

____

Name ____________________

LESSON 1.2

Benchmark Numbers

Vocabulary

Fill in the blank.

1. A ____________________ is a point of reference.

Use the benchmark number to choose the more reasonable estimate.

2. 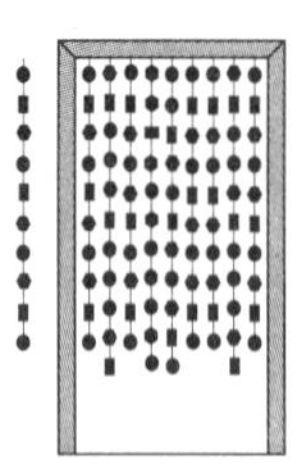

90 or 900 ________

3. 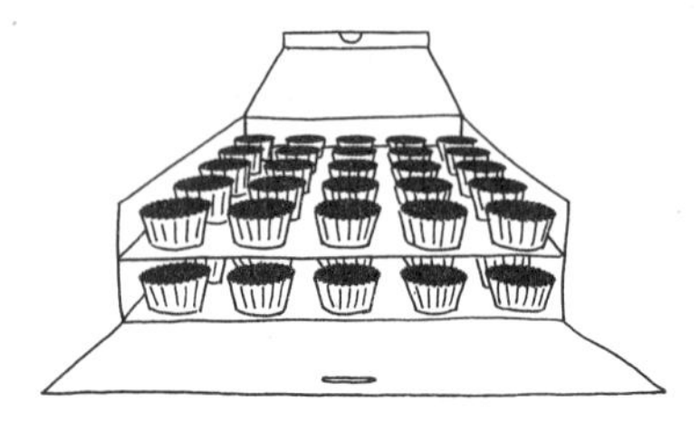

50 or 500 ________

Write *yes* or *no* to tell whether the estimate is reasonable.

4. • 40 nickels in a roll
• 8 rolls of nickels
Estimate: There are about 300 nickels.

____

5. • 8 slices per pizza
• 20 pizzas in the cafeteria
Estimate: There are about 100 pizza slices in the cafeteria.

____

6. • 20 tables in a restaurant
• Each table seats 6 people.
Estimate: The restaurant seats about 100 people.

____

7. • 10 rolls of film in a case
• Each roll has 24 exposures.
Estimate: There are about 100 exposures of film in a case.

____

Mixed Applications

8. Each shelf in a store can display about 30 boxes of cards. The store has 11 shelves. About how many boxes of cards can the store display?

____

9. Each of 65 students drank a fruit juice for lunch. If the fruit juice comes in six-packs, how many six-packs were needed?

____

Name ____________________

LESSON 1.3

Place Value to Hundred Thousands

Write the value of the **bold-faced** digit.

1. 2**5**1,469 ____________ ____________

2. 747,**3**22 ____________ ____________

3. 63**8**,415 ____________ ____________

Complete.

4. 371,695 = (____ × 100,000) + (____ × 10,000) + (____ × 1,000) + (____ × 100) + (____ × 10) + (____ × 1)

5. 279,956 = (____ × 100,000) + (____ × 10,000) + (____ × 1,000) + (____ × 100) + (____ × 10) + (____ × 1)

6. 986,741 = (9 × ________) + (8 × ________) + (6 × ________) + (7 × ________) + (4 × ________) + (1 × ________)

Write the expanded numbers in standard form.

7. 400,000 + 70,000 + 9,000 + 500 + 60 + 9 ____________

8. 100,000 + 90,000 + 0 + 300 + 0 + 1 ____________

9. 700,000 + 50,000 + 5,000 + 0 + 40 + 6 ____________

10. 600,000 + 80,000 + 3,000 + 0 + 0 + 0 ____________

Mixed Applications

11. In 1994, the American Kennel Club registered 132,051 Labrador Retrievers. What is the value of the digit 3 in this number?

____

____

12. This year, Connie earned $56,426 as a sales manager. Last year she earned $49,378. How much more did Connie earn this year?

____

Name ______________________________

LESSON 1.4

Place Value to Hundred Millions

Write the value of the **bold-faced** digit.

1. 301,**7**24,469
2. 4**2**6,379,831
3. 234,5**6**7,890
4. **1**89,612,357
5. 52**1**,874,394
6. 879,46**4**,978

Write two other forms for each number.

7. 125,740,689
8. 200,403,926
9. 600,000,000 + 0 + 9,000,000 + 700,000 + 60,000 + 0 + 400 + 30 + 1
10. eight hundred ninety-five million, four hundred six thousand, two hundred twenty-five

Mixed Applications

11. The CD-ROM for *Grolier's Multimedia Encyclopedia* holds 638,400,000 bits of information. What is the value of the digit 8 in this number?

12. Every 225 Earth days, the planet Venus travels around the sun an average distance of 67,000,000 miles. Write the distance in word form.

Name ______________________________

Comparing and Ordering

Write <, >, or = for each ◯ .

1. 532 ◯ 523

2. 6,475 ◯ 6,574

3. 7,821 ◯ 7,821

4. 59,495 ◯ 54,945

Order from greatest to least.

5. 354; 345; 435

____

6. 7,685; 7,856; 7,568

____

7. 45,732; 47,325; 42,537

____

8. 654,837; 675,384; 648,753

____

Order from least to greatest.

9. 647; 746; 467

____

10. 6,785; 6,875; 6,587

____

11. 15,234; 15,423; 15,342

____

12. 324,987; 432,789; 734,298

____

Mixed Applications

For Problems 13–16, use the table.

Waterfall	Length
Victoria Falls	355 ft
Yellowstone Falls	308 ft
Churchill Falls	250 ft
Iguacu Falls	280 ft

13. Which of these waterfalls is the longest?

____

14. Which waterfall is longer, Iguacu Falls or Churchill Falls?

____

15. How much longer is Victoria Falls than Churchill Falls?

____

16. Which of these waterfalls is the shortest?

____

Name ______________________________

LESSON 1.5

Problem-Solving Strategy

Use a Table

Use a table and solve.

1. This table shows some of the Hawaiian Islands and their areas. Which island has the greatest area? the least area?

____

HAWAIIAN ISLANDS	
Island	**Area (square miles)**
Maui	729
Hawai'i	4,038
O'ahu	608
Kaua'i	553

2. This table shows some of the major cities in California and their populations. Order the cities from largest population to smallest population.

____

____

CALIFORNIA CITIES	
City	**Population**
San Diego	1,110,549
Los Angeles	3,485,398
Long Beach	429,433
San Francisco	723,959

Mixed Applications

Solve.

CHOOSE A STRATEGY

- **Work Backward**
- **Use a Table**
- **Act It Out**
- **Make an Organized List**

3. Romona worked 9 hours over the weekend. She worked twice as many hours on Saturday as on Sunday. How many hours did she work on Saturday? on Sunday?

____

4. Anna spent half her money on lunch. She spent half her remaining money on a book. She had $2 remaining. How much money did Anna begin with?

____

5. School T-shirts are available in sizes small, medium, and large. Color choices are blue or red. Make a list of all possible T-shirt choices.

____

____

6. Allison is shorter than Pam and taller than Jane. Jane is taller than Ellen. Who is the tallest person?

____

Name ___________________________________

Adding and Subtracting with Data

Vocabulary

Fill in the blank.

1. Addition and subtraction are ___________________________.
   One operation undoes the other operation.

For Exercises 2–4, use the data in the table.

Sixth-Grade Book Collection	
Mr. Tyse	72 books
Ms. Carvajal	86 books
Mrs. Langway	91 books
Ms. Beebe	81 books
Mrs. Voglino	78 books
Ms. Benfosado	83 books

2. How many books were collected in all?

   ____

3. How many more books were collected by Ms. Carvajal's class than Mr. Tyse's class?

   ____

4. Which two classrooms together collected 159 books?

   ____

Write the related number sentence. Find the sum or difference.

5. $71 - 53 = n$

   ____

6. $114 - 86 = n$

   ____

7. $63 + n = 91$

   ____

8. $36 + n = 68$

   ____

Mixed Applications

9. Niki scored 36 total points in a basketball game. She scored 12 points in the second half of the game. How many points did Niki score in the first half of the game?

   ____

10. Niki's basketball team won 12 games and lost 18 games. How many games did her team play?

    ____

Name ________________________

LESSON 2.2

More About Subtracting

Draw the counters after regrouping. Find the difference.

1. 635 − 327

Hundreds	Tens	Ones
○○○ ○○○	○○ ○	○○ ○○ ○

Hundreds	Tens	Ones

2. 514 − 227

Hundreds	Tens	Ones
○○○ ○○	○	○○ ○○

Hundreds	Tens	Ones

Find the difference. You may use counters.

3. 321 − 174

4. 476 − 298

5. 512 − 251

6. 347 − 165

7. 722 − 457

8. 657 − 278

9. 928 − 439

10. 845 − 259

11. 1,452 − 694

12. 3,425 − 2,857

13. 5,413 − 2,336

14. 8,641 − 3,568

Mixed Applications

For Problems 15–17, use the table.

BASKETBALL LEADING SCORERS	
Name	**Points**
Jeff	342
Bill	276
Dave	245

15. How many more points did Jeff score than Bill?

____

16. How many more points did Jeff score than Dave?

____

17. What is the total number of points scored by Jeff, Bill, and Dave?

____

18. Nadia bought 5 books that cost $3.99 each. What was the total cost of Nadia's purchase?

____

Name ______________________

LESSON 2.3

Subtracting Across Zeros

Show how you regrouped for each problem. Solve.

1. 500 − 236

2. 3,000 − 1,721

3. 7,003 − 4,875

4. 15,000 − 11,484

Find the difference.

5. 400 − 265

6. 700 − 527

7. 3,000 − 782

8. 4,003 − 2,726

9. 9,000 − 7,623

10. 5,000 − 2,966

11. 10,000 − 8,329

12. 14,000 − 6,451

13. 16,000 − 11,469

14. 20,000 − 6,817

15. 25,000 − 13,475

16. 60,000 − 28,937

Mixed Applications

17. Hillside Elementary School has 625 students. Valley Elementary School has 438 students. How many more students go to Hillside?

____

18. The Great Western Forum seats 17,505 fans for basketball games. If there are 2,638 empty seats, how many people are at the game?

____

19. A writer took in $6,001. She paid $2,016 in taxes. How much money did she make?

____

20. There are 2,000 milliliters in a 2-liter bottle of soda. If you serve five 250 milliliter cups of soda, how many milliliters of soda do you have left?

____

Name ______________________

LESSON 2.4

Choosing Addition or Subtraction

Choose and name the operation. Solve.

1. Mimi collected 246 bottles to recycle. Carlos collected 178 bottles. How many more bottles did Mimi collect?

____

2. Brett had 428 animal stickers. He received 35 more for his birthday. What is the total number of stickers Brett has?

____

3. The school raffle committee printed 15,000 raffle tickets. The students sold all but 924. How many tickets did the students sell?

____

4. George ate a sandwich with 350 calories. He drank a glass of milk with 110 calories. How many total calories did George consume?

____

5. Gretchen was born in 1990. Her grandfather was born in 1933. How much older than Gretchen is her grandfather?

____

6. Katrina bought a CD player for $126 and 3 CD's for $29. How much did she spend in all?

____

Mixed Applications

For Problems 7–10, use the table.

THE GREAT LAKES		
Lake	Length	Maximum Depth
Superior	350 mi	1,330 ft
Michigan	307 mi	923 ft
Erie	241 mi	210 ft
Huron	206 mi	750 ft
Ontario	193 mi	802 ft

7. Is the total length of the Great Lakes greater than or less than 1,200 miles?

____

8. Lake Erie is the shallowest Great Lake. How much deeper is Lake Superior?

____

9. The Illinois River is 240 mi long. How much longer is Lake Michigan than the Illinois River?

____

10. What is the total depth of Lake Erie and Lake Michigan?

____

Name ____________________

LESSON 2.5

Estimation and Column Addition

Vocabulary

Fill in the blank.

1. Numbers that are easy to compute mentally are

____________________ .

Choose a method, and then estimate each sum.

2.
```
   17
   26
   34
 + 42
```

3.
```
   28
   45
   62
 + 76
```

4.
```
   91
   16
   25
 + 34
```

5.
```
   18
   74
   69
 + 86
```

6.
```
   118
   235
   486
 + 379
```

7.
```
   449
   250
   368
 + 196
```

8.
```
   751
   832
   244
 + 628
```

9.
```
   340
   568
   455
 + 832
```

10.
```
   647
   353
   621
 + 550
```

11.
```
   191
   157
   438
 + 571
```

12.
```
   361
   158
   311
 + 769
```

13.
```
   261
   342
   557
 + 923
```

Mixed Applications

14. In Mrs. Ryan's class, Mike, Bill, Sharim, and Jo collected cans for recycling. Mike collected 139 cans, Bill—125 cans, Sharim—159 cans, and Jo—246 cans. About how many cans were collected by all four students?

____

15. In a timed situp test, Mike did 42 situps, Bill did 24 situps, Sharim did 31 situps, and Jo did 48 situps. How many situps did they do in all?

____

Name ___________________

LESSON 2.5

Problem-Solving Strategy

Estimate or Exact Answer?

Decide whether you need to estimate, find the exact answer, or both. Solve.

1. Ben received $10.00 for doing chores. He wants to buy some cards for $2.89, an action figure for $4.99, and a comic book for $1.79. Does he have enough to pay for all three items? How much will Ben pay at the cash register? How much change will he receive?

   ____

2. Yasmin received $50.00 for her birthday. She wants to buy a sweater for $13.99, a necklace for $14.95, and shoes for $19.98. Does Yasmin have enough money to pay for all three items? How much will she pay? How much change will she receive?

   ____

3. Catherine wants to buy some roses for $6.99, some potting soil for $3.98, and a ceramic pot for $7.95. She has $20.00. How much will she spend on the items? How much change will she receive?

   ____

4. Mark has $30.00 to spend on party favors. He wants to buy party horns for $9.95, candy for $5.98, and magic tricks for $15.99. Does Mark have enough money for all three items?

   ____

Mixed Applications

Solve.

CHOOSE A STRATEGY

- Use a Table
- Guess and Check
- Write a Number Sentence
- Work Backward

5. Walt bought a CD player on sale for $99.95 plus $4.99 tax. The regular price was $149.99 including tax. How much did Walt save?

   ____

6. Emma spent $4 on cards and $18 on a sweater. Emma has $9 left. How much did Emma begin with?

   ____

7. In an even 2-digit number, the second digit is 3 times the first. What is the number?

   ____

8. Don is a cashier. When he calculates the amount of change, does he want an estimate or the exact answer?

   ____

Name ____________________

Using Tenths and Hundredths

Write the letter of the decimal that matches each model.

1. 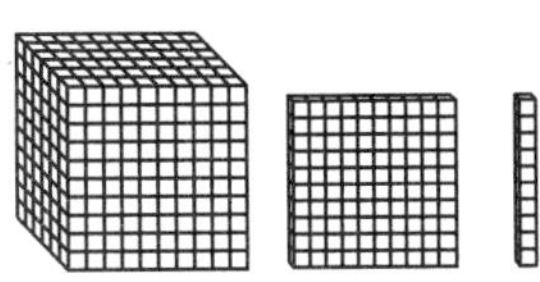
____

2. 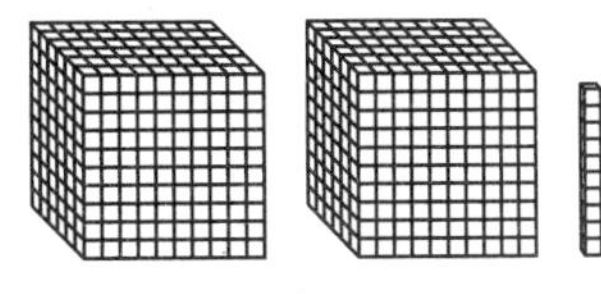
____

3. 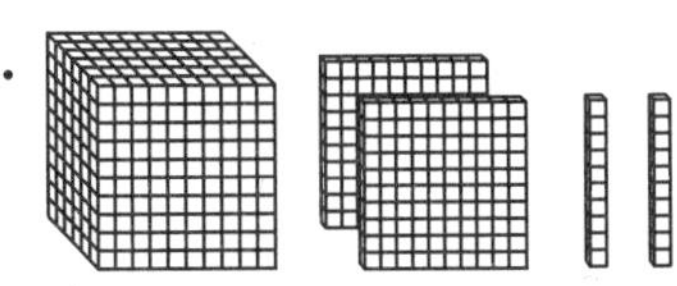
____

a. 2.03 b. 1.22 c. 1.11

Write the decimal for each.

4. eight and thirteen hundredths ____________

5. four and three tenths ____________

6. sixty-five hundredths ____________

7. five and one tenth ____________

8. six and five hundredths ____________

9. seven and twenty-two hundredths ____________

10. 3 + 0.4 ________

11. 5 + 0 + 0.01 ________

12. 7 + 0.9 + 0.03 ________

13. 0.2 + 0.09 ________

14. 4 + 0.5 + 0.02 ________

15. 12 + 0.7 + 0.04 ________

Mixed Applications

16. Sophia ran the 100-meter dash in 11.36 seconds. What is the value of the 3 in her time?
____

17. Frank ran the 100-meter race in 10.09 seconds. Write Frank's time in written form.
____

18. One school district raised $12,195 in a town-wide fund raiser. A second school district raised $9,468. How much more money did the first district raise?
____

19. Gloria's room is 12 feet long and 15 feet wide. What is the area of Gloria's room? the perimeter?
____
____

Name ______________________________

LESSON 3.2

Thousandths

Vocabulary

Fill in the blank.

1. One milliliter equals one ______________ of a liter.

Use base-ten blocks to model each number. Record the number in a place-value chart.

2. 6.254

Ones	Tenths	Hundredths	Thousandths
.			

3. 9.064

Ones	Tenths	Hundredths	Thousandths
.			

Write in standard form.

4. six thousandths

____

5. sixty-eight thousandths

____

6. four hundred twenty-two thousandths

____

7. three and six hundred twelve thousandths

____

Write in written form.

8. 0.009

____

9. 0.067

____

10. 0.723

____

____

11. 3.614

____

____

Mixed Applications

12. Jeremy lifts a 12-lb barbell 8 times per set. He does 3 sets. How many total pounds does Jeremy lift?

____

13. Izeta's time for the 100-meter dash was 11.824 seconds. In what place-value position is the digit 4?

____

Name ___________________________

LESSON 3.3

Place Value

Write in standard form.

1. 5,000 + 400 + 30 + 7 + 0.6 + 0.01 + 0.009

____

2. six thousand, seven hundred six and four hundred thirty-five thousandths

____

3. 7,000 + 300 + 0 + 0 + 0.1 + 0.07 + 0.002

____

4. two thousand, seventy seven and eight hundred six thousandths

____

Write the value of the digit 6 in each number.

5. 3,628.314

____

6. 4,921.567

____

7. 7,847.126

____

8. 9,761.478

____

Mixed Applications

For Problems 9–11, use the table.

1-MILE RUN	
Runner	**Time in seconds**
Harris	237.050
Armstrong	237.255
Albaro	240.005

9. Write Albaro's time in written form.

____

____

10. Write Armstrong's time in expanded form.

____

____

11. Who won the 1-mile run? Who placed second?

____

____

Name ______________________________

LESSON 3.4

Equivalent Decimals

Vocabulary

Fill in the blank.

1. ____________ ____________ are different names for the same number or amount.

Write *equivalent* or *not equivalent* to describe each set of decimals.

2. 0.4 and .40

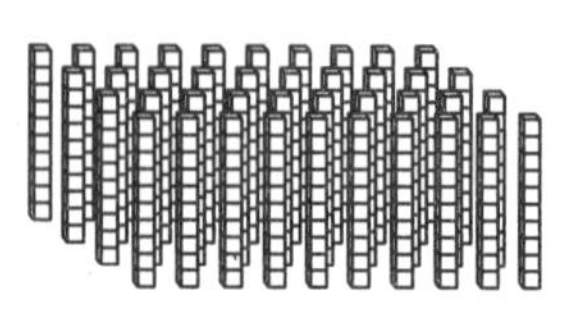

____

3. 0.08 and 0.008

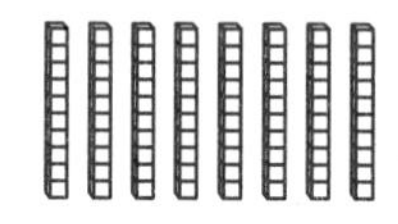

____

Write an equivalent decimal for each.

4. 1.2 ____________

5. 3.71 ____________

6. 0.060 ____________

7. 6.200 ____________

8. 3.450 ____________

9. 4.15 ____________

10. 2.4 ____________

11. 7.30 ____________

12. 2.900 ____________

Mixed Applications

13. Vinny won the 400-meter race in 48.69 seconds. Write an equivalent decimal for his time.

____

14. Jami's family drove 175.6 miles to a vacation spot. Write an equivalent decimal for the distance they traveled.

____

15. Dennis had 9 rebounds in each of his team's 15 games. What was the total number of rebounds for Dennis?

____

16. What number has 8 in the thousand and thousandths place, 7 in the hundreds and hundredths place, 4 in the tens and tenths place, and 0 in the ones place?

____

Name ______________________

Comparing and Ordering

Use the number line to compare the decimals.

3.0 3.1 3.2 3.3 3.4 3.5 3.6 3.7 3.8 3.9 4.0

1. 3.62 ◯ 3.26
2. 3.20 ◯ 3.02
3. 3.59 ◯ 3.6
4. 3.30 ◯ 3.4
5. 3.54 ◯ 3.45
6. 3.15 ◯ 3.25

Write <, >, or = in each ◯.

7. 0.25 ◯ 0.23
8. 6.56 ◯ 6.65
9. 7.21 ◯ 7.210
10. 27.35 ◯ 27.53
11. 368.58 ◯ 368.85
12. 237.524 ◯ 237.254
13. 736.54 ◯ 736.540
14. 16.2 ◯ 16.200
15. 878.787 ◯ 878.878

Order from least to greatest.

16. 7.11, 7.09, 7.07 ______
17. 12.54, 12.45, 12.65 ______
18. 3.020, 3.002, 3.200 ______
19. 17.560, 17.065, 17.056 ______
20. 2.654, 2.546, 2.456, 2.465 ______

Mixed Applications

For Problems 21–23, use the menu.

Sandwich Menu	
Turkey	$2.65
Tuna	$2.95
Chicken	$3.25
Ham	$2.75
Cheese	$2.25

21. Which sandwiches cost more than turkey but less than chicken?

______

22. What is the cost of a ham and a tuna sandwich?

______

23. Write the sandwiches in order from the least expensive to the most expensive.

______

24. The swim team had $500 to spend on equipment. After spending $367, how much does the team have left to spend?

______

Name ______________________________

LESSON 3.5

Problem-Solving Strategy

Make a Table

Make a table and solve.

1. Three friends compared their batting averages. Jane batted 0.321, Nick batted 0.231, and Fred batted 0.312. Who had the highest average?

   ____

2. On a hot summer day in Arizona, the temperature was 115.6°F in Scottsdale, 116.5°F in Phoenix, and 115.9°F in Sun City. Where was the highest temperature?

   ____

3. The local deli offers three sandwiches. Tuna costs $3.25, turkey costs $2.95, and ham costs $3.35. Which sandwich is the most expensive?

   ____

4. Three students were in the high-jump competition. Mark jumped 1.54 meters, Peter jumped 1.45 meters, and Nader jumped 1.61 meters. Who made the highest jump?

   ____

Mixed Applications

Solve.

CHOOSE A STRATEGY

- Use a Table
- Work Backward
- Draw a Diagram
- Write a Number Sentence
- Guess and Check

5. Fredrica's car gets 24 miles to a gallon of gasoline. Its tank holds 19 gallons of gasoline. How far can the car be driven on one tank of gas?

   ____

6. Five students stand in a line. Ned is in the middle. Bart stands next to Ned. Lisa is at the right end. Matt stands between Ned and Lisa. Where is Margie?

   ____

7. Terri scored twice as many points in the second half of a basketball game than in the first half. She scored 24 second-half points. How many total points did Terri score?

   ____

8. Last year in Detroit it snowed 21.8 cm in December, 17.4 cm in January, 41.3 cm in February, and 24.9 cm in March. In which month did the most snow fall?

   ____

Name ______________________________

Adding Decimals

Use base-ten blocks to model. Record the sum on a place-value chart.

1. 0.2 + 0.9

2. 0.23 + 0.35

3. 1.251 + 0.833

Find the sum.

4. 9.2 + 0.8

5. 7.2 + 3.6

6. 8.94 + 2.06

7. 1.257 + 3.333

8. 1.125 + 5.725

9. 2.4 + 1.5

10. 3.2 + 5.6

11. 1.38 + 0.91

12. 1.523 + 0.374

13. 3.682 + 2.479

14. 3.2 + 1.4 = ________

15. 2.68 + 3.14 = ________

16. 4.098 + 1.875 = ________

17. 5.7 + 3.6 = ________

18. 8.05 + 3.52 = ________

19. 1.111 + 2.222 = ________

Mixed Applications

20. Greg went to the store. He bought a bag of raisins for $1.29 and a carton of juice for $0.85. How much did he spend?

____

21. Susan found $0.63 in change. Her brother found $0.57. Together how much money did they find?

____

22. The art teacher Mr. Terry bought 25 blue pencils, 35 red pencils, and 48 green pencils. How many pencils did he buy in all?

____

23. Brad's family traveled 250 miles on Monday and 345 miles on Tuesday. Grace's family traveled 187 miles on Monday and 432 miles on Tuesday. Which family traveled more miles?

____

Name ______________________

LESSON 4.2

More About Adding Decimals

Use an equivalent decimal to find the sum.

1. 1.24 + 0.317 = ______
2. 2.082 + 3.6 = ______
3. 1.38 + 3.849 = ______
4. 2.54 + 3 = ______
5. 35.7 + 12.426 = ______
6. 37.84 + 6.592 = ______
7. 3.12 + 6 = ______
8. 21.3 + 15.264 = ______
9. 29.04 + 9.846 = ______

10. 7 + 0.3
11. 4.6 + 1.54
12. 4.06 + 0.873
13. 8 + 4.75
14. 6.344 + 0.22

15. 5 + 0.6
16. 3.9 + 2.61
17. 7.04 + 0.984
18. 6 + 5.267
19. 2.841 + 0.13

20. 2.5 + 3.47
21. 3.1 + 2.73
22. 8.4 + 5.283
23. 7.683 + 2.3
24. 13.2 + 4.08

25. 7.03 + 4.2
26. 8.35 + 0.294
27. 9.2 + 8.416
28. 4.257 + 5.5
29. 21.3 + 6.42

Mixed Applications

30. Tony rides his bike in the mountains. He writes how far he rides each day in his weekly log. How many miles did he ride this week?

______

Weekly Biking Log				
Mon	Tue	Wed	Thur	Fri
2.5 mi	1.6 mi	3.7 mi	1.8 mi	2.1 mi

31. Samantha bought a new pair of ski gloves for $4.95 and 2 scarves for $2.78 each. How much did she spend for all her purchases?

______

32. Roger sold 258 tickets to the school play. Fredrica sold 392 tickets. How many more tickets did Fredrica sell than Roger?

______

Name ______________________

Subtracting Decimals

Find the difference.

1. $2.5 - 0.8$
2. $3.4 - 3.1$
3. $2.04 - 1.7$
4. $3.6 - 2.7$
5. $3.5 - 1.04$
6. $1.6 - 0.8$
7. $4.8 - 4.2$
8. $3.07 - 2.8$
9. $4.2 - 3.8$
10. $6.7 - 2.02$
11. $3.87 - 1.36$
12. $2.7 - 1.824$
13. $5.426 - 2.56$
14. $12.537 - 4.315$
15. $19.469 - 12.253$
16. $4.68 - 2.15$
17. $3.2 - 2.451$
18. $7.264 - 3.49$
19. $16.852 - 8.231$
20. $17.578 - 13.154$

21. $2.36 - 1.17 =$ ______
22. $1.7 - 0.763 =$ ______
23. $2.8 - 1.9 =$ ______
24. $3.75 - 2.68 =$ ______
25. $2.4 - 1.468 =$ ______
26. $3.1 - 2.5 =$ ______
27. $3.68 - 1.892 =$ ______
28. $5.2 - 3.181 =$ ______
29. $6.42 - 3.374 =$ ______
30. $4.21 - 2.362 =$ ______
31. $7.3 - 4.226 =$ ______
32. $5.69 - 2.473 =$ ______

Mixed Applications

For Problems 33–35, use the table.

Speeds of Animals	
Quarter horse	47.5 mph
Greyhound	39.35 mph
Human	27.89 mph
Snail	0.03 mph

33. The maximum speeds of animals over one-quarter mile varies greatly. What is the difference between the fastest and the slowest animal?

____

34. How much faster is a greyhound than a human?

____

35. In the snail's speed, what is the place value of the 3?

____

Name ______________________________

Estimating Sums and Differences

Estimate the sum or difference to the nearest tenth.

1. 6.45 − 2.81
2. 7.32 − 5.14
3. 7.68 + 3.52
4. 18.07 + 11.66
5. 27.36 − 15.04

Estimate the sum or difference to the nearest hundredth.

6. 1.285 + 0.822
7. 2.843 + 7.158
8. 4.060 − 3.724
9. 6.341 − 1.636
10. 2.578 − 0.372

Estimate the sum or difference and compare. Write < or > in each ◯.

11. 7.21 − 5.56 ◯ 6.89 − 2.34
12. 4.73 + 3.29 ◯ 5.32 + 2.99
13. 9.213 + 4.764 ◯ 8.345 + 6.754
14. 36.84 − 15.49 ◯ 58.94 − 37.99
15. 45.76 + 21.84 ◯ 32.98 + 34.05
16. 52.85 + 34.76 ◯ 46.34 + 39.82
17. 9.034 − 4.571 ◯ 7.562 − 2.199
18. 6.045 − 2.374 ◯ 8.461 − 5.921

Mixed Applications

For Problems 19–21, use the table.

19. Value-Mart and Save-More both advertise that they have the lowest prices in town. Which store offers the best price on pretzels? How much less is the cost per ounce?

____

COMPARISON SHOPPING		
	Value-Mart	Save-More
Pretzels	0.326¢ per ounce	0.33¢ per ounce
Cheese	$1.095 per pound	$1.113 per pound

20. Maria and Sara compared the price of cheese. In which store is the cheese more expensive? How much more expensive per pound is it?

____

21. Sara spent $3.14 on pretzels and $5.89 on cheese. About how much did Sara spend in all?

____

Name ____________________

Choosing Addition or Subtraction

Choose and name the operation. Solve.

1. Grace and Shelly are practicing for the marathon. Grace runs 23.8 miles and Shelly runs 19.9 miles. How many more miles does Grace run than Shelly?

____

2. Jim buys two slot cars and tests them on a local track. The blue car has a lap time of 6.987 seconds and the red car has a lap time of 8.732 seconds. Which car is faster? How much faster?

____

3. Jim is visiting his grandparents. It is 85.7 miles from his house to their house. If he drives there and back, how many miles does he travel?

____

4. Tony buys a collector's set of baseball cards for $32.75. His mother buys him a set for $45.98. How much money do he and his mother spend on baseball cards?

____

For Problems 5–7, use the items shown below.

5. Sue went shopping at the annual clearance sale. She bought a shirt and 2 pairs of socks. How much money did Sue spend?

____

6. Sue paid for her clothes with a $20 bill. How much change did she receive?

____

7. Tom bought a hat. His change was $5.24. How much money did he have to start with?

____

Mixed Applications

8. Sal and Alice planted trees for the Forestry Service. Last weekend Sal planted 113 trees and Alice planted 97 trees. How many more trees did Sal plant?

____

9. Cheryl wants to put a border around her window. The window is $3\frac{1}{2}$ feet wide and $5\frac{1}{4}$ feet high. How much border does she need to go around the window?

____

Name ____________________

Problem-Solving Strategy

Write a Number Sentence

Write number sentences to solve.

1. Devin's checking account has a starting balance of $78.15. He writes checks for $15.78, $23.89, and $5.10. He makes 1 deposit of $68.00. What is his balance? Does he have enough to write a check for $90.00?

2. Lisa's bank statement shows a balance of $23.75. The statement shows deposits of $67.35 and $150.00. She writes checks for $55.00 and $62.88. What is Lisa's current balance?

3. Ingrid can run a 100-meter dash in 10.49 seconds. Will her team beat the 400-meter record of 41.37 seconds if her 3 teammates also run 100 meters in 10.49 seconds?

4. At the paper airplane derby, the team from Flight Town U.S.A. has hang times of 11.7 seconds, 9.6 seconds, 10.5 seconds, and 12.3 seconds. What is the team's total hang time?

Mixed Applications

Solve.

CHOOSE A STRATEGY

- Use a Table • Draw a Diagram • Guess and Check • Work Backward • Write a Number Sentence

For Problems 5–6, use the table.

Weekly Miles Driven	
Week 1	1,832
Week 2	2,164
Week 3	1,624
Week 4	857

5. Tina traveled with her father in his truck. His truck travels about 450 miles on 1 tank of gas. About how many tanks of gas did they use in the fourth week?

6. Tina spends all but the third week traveling with her father. How many miles does Tina ride with her father during the month?

7. Did Tina and her father drive more in weeks 1 and 2 or in weeks 3 and 4? How much more?

Name ______________________________

Using Multiplication Properties

Vocabulary

Write the correct letter from Column 2.

Column 1

___ **1.** Zero Property for Multiplication

___ **2.** Property of One

___ **3.** Associative Property of Multiplication

___ **4.** Commutative Property of Multiplication

Column 2

a. When one of the factors is 1, the product equals the other number.

b. You can group factors differently. The product is always the same.

c. When one factor is 0, the product is 0.

d. You can multiply numbers in any order. The product is always the same.

Write the name of the multiplication property used in each number sentence.

5. $2 \times 6 = 6 \times 2$

____

____

6. $(4 \times 2) \times 5 = 4 \times (2 \times 5)$

____

____

7. $7 \times 1 = 7$

____

____

Complete the equation. Identify the property used.

8. $3 \times \square = 0$

____

____

9. $\square \times 7 = 7 \times 6$

____

____

10. $(5 \times 4) \times 3 = 5 \times (\square \times 3)$

____

____

Mixed Applications

11. It takes Mel 23 minutes to walk to school. It takes Julia 17 minutes to walk to school. How much longer does it take Mel to get to school than Julia?

____

12. In Mr. Pod's class, there are 6 desks with 3 students at each desk. In Ms. Carter's class, there are 7 desks with 2 students at each desk. Which classroom has more students?

____

Name ____________________

LESSON 5.2

Recording Multiplication

Explain how you would model with colored counters.

1. $\begin{array}{r} 304 \\ \times \quad 4 \\ \hline \end{array}$

2. $\begin{array}{r} 461 \\ \times \quad 5 \\ \hline \end{array}$

3. $\begin{array}{r} 200 \\ \times \quad 6 \\ \hline \end{array}$

4. $\begin{array}{r} 705 \\ \times \quad 2 \\ \hline \end{array}$

Solve by using colored counters.

5. $2 \times 156 = n$ ____________

6. $3 \times 450 = n$ ____________

7. $4 \times 392 = n$ ____________

8. $6 \times 378 = n$ ____________

Mixed Applications

For Problem 11, use the table.

9. Frank and Felicia arrived at the restaurant at 7:00 P.M. They stayed for 1 hour. It took them 10 minutes to get home, 15 minutes to pay the baby-sitter, and 30 minutes to get the kids to bed. What time were the kids in bed?

____

10. Sam rode his bike to school each day for 29 days. The total distance each day was 5 miles. Ariel rode her bike to school each day for 20 days. The total distance each day was 8 miles. Who rode a greater distance?

____

11. Tamara and David went out to lunch on Thursday. They each bought a slice of pizza, a salad, and a drink. How much did they spend in all?

____

Lunch Prices	
Pizza	$1.25
Salad	$1.75
Drink	$1.10
Chips	$0.70

Name ______________________

LESSON 5.3

Practicing Multiplication

Find the product.

1. 315×3 ______

2. 421×8 ______

3. 526×9 ______

4. 137×2 ______

5. 213×6 ______

6. 336×4 ______

7. 554×7 ______

8. 692×5 ______

9. $428 \times 8 = n$ ______

10. $361 \times 3 = n$ ______

11. $603 \times 5 = n$ ______

12. $830 \times 9 = n$ ______

13. $531 \times 6 = n$ ______

14. $222 \times 4 = n$ ______

15. $651 \times 6 = n$ ______

16. $108 \times 9 = n$ ______

17. $752 \times 4 = n$ ______

18. $903 \times 6 = n$ ______

19. $362 \times 4 = n$ ______

20. $841 \times 2 = n$ ______

Mixed Applications

21. Janet and Alan have 236 stamps. They collect 20 more each day for 3 days. What is the total amount of stamps they have at the end of 3 days?

______

22. Tomika had $15.23. She went to the store and bought an eraser, notebook, stapler, and pencils for $12.35. How much change did she get back?

______

23. Suzie plays basketball. Her team's total for the season is 409 points. They scored the following points in their first 4 games: 80, 85, 76, and 83. How many points did they score in their fifth game?

______

24. Jacob does 125 sit-ups each day. He does them for a week. How many sit-ups does he do in one week?

______

Name ____________________

LESSON 5.4

Finding Volume

Vocabulary

Fill in the blanks.

1. The measure of the space a solid figure occupies is ____________.

2. Volume is measured in ____________.

Tell how many unit cubes were used to build each rectangular prism. Find the volume.

3. 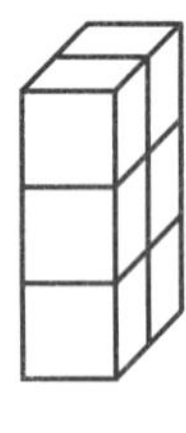

units ____________

volume ____________

4. 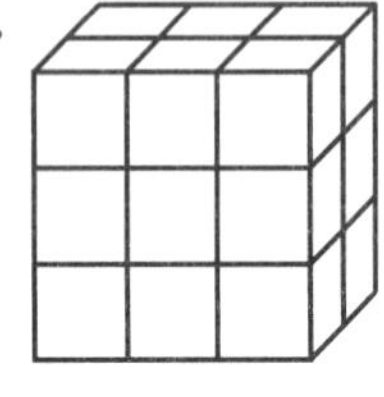

units ____________

volume ____________

5. 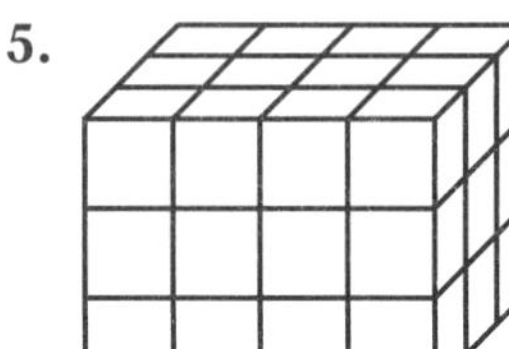

units ____________

volume ____________

6. 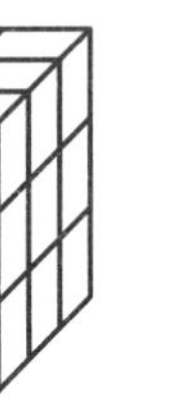

units ____________

volume ____________

Use unit cubes to build each prism. Complete the table.

	Length of Base	Width of Base	Height (number of layers)	Volume (cubic units)
7.	7 cubes	3 cubes	5 cubes	____
8.	6 cubes	2 cubes	4 cubes	____
9.	8 cubes	4 cubes	3 cubes	____

Mixed Applications

10. Shana has two shoe boxes. One shoe box is $4 \times 3 \times 5$ cubic units. The other shoe box is $3 \times 4 \times 6$ cubic units. Which shoe box has the larger volume?

____

11. Tim receives $1.25 for every piece of trash he picks up. He picks up 9 pieces on Monday and 8 pieces on Tuesday. How much money does Tim earn?

____

Name ______________________________

LESSON 5.5

Finding Area and Volume

Vocabulary

Fill in the blank.

1. A ______________ is a set of symbols that expresses a mathematical rule.

The formula for area is Area = length × width, or $A = l \times w$.

The formula for volume is Volume = length × width × height, or $V = l \times w \times h$.

Find the area.

2. 6 in. 10 in.

____

3. $l = 36$ cm

w = 4 cm

$A = \square$ sq cm

____

4. $l = 16$ m

w = 3 m

$A = \square$ sq m

____

Find the volume.

5.

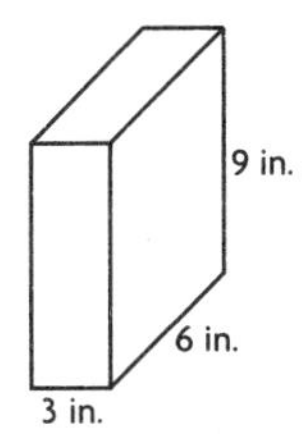

____

6. $l = 5$ cm

w = 7 cm

$h = 8$ cm

$v = \square$ cu cm

____

7. $l = 4$ in.

w = 6 in.

$h = 8$ in.

$v = \square$ cu in.

____

Mixed Applications

8. There are 22 students in Mr. Petri's fourth grade class. Each student has 4 textbooks. How many textbooks are there in all?

____

____

____

9. Tom's shelf is 6 feet long, 4 feet wide, and 3 feet high. Kim's shelf is 7 feet long, 4 feet wide, and 4 feet high. Who has the shelf with the larger volume? Explain.

____

____

____

Name ______________________________

LESSON 5.5

Use a Formula

Use a formula to solve.

1. Keesha is painting a mural. The wall is 11 feet long and 8 feet wide. What is the area of the wall that Keesha is painting?

   ____

2. The Carreys are fencing in their backyard. The fence measures 18 feet long and 9 feet wide. What is the area of their backyard?

   ____

3. Chelsey sees two doghouses to buy. One measures 4 ft × 4 ft × 3 ft. The other measures 3 ft × 5 ft × 5 ft. She wants to get the one with the larger volume. Which doghouse should she buy?

   ____

4. Ron likes two dressers with the same features. One measures 4 ft × 4 ft × 2 ft and costs $76. The other measures 3 ft × 5 ft × 2 ft and costs $85. Which dresser is the better buy?

   ____

Mixed Applications

Solve.

CHOOSE A STRATEGY

- **Draw a Diagram**
- **Write a Number Sentence**
- **Use a Table**
- **Make a Model**

5. Kim walks 3 miles to the playground and 3 miles home. She walks there and back 4 days a week. How many miles does she walk in a week?

   ____

6. Hank's vegetable garden measures 8 feet long and 6 feet wide. He starts at one corner and plants marigolds along the edge. The marigolds are 1 foot apart. How many marigolds does he plant?

   ____

7. Jeanette has $52.00. She wants to buy a coat for $35.69, a scarf for $8.99, and a pair of gloves for $5.50. What is the total cost for all those items? Does she have enough money?

   ____

   ____

   ____

8. Lauren wants to buy an aquarium that has a volume of 18 cubic units. How many different aquariums with the same volume could she buy? Write the measurements for each one.

   ____

   ____

   ____

Name ______________________________

Using the Distributive Property

Vocabulary

Fill in the blanks.

1. The ______________ ______________ allows you to break apart numbers to make them easy to multiply.

Use the grid below to model. Find the product.

2. $10 \times 17 = n$

3. $15 \times 14 = n$

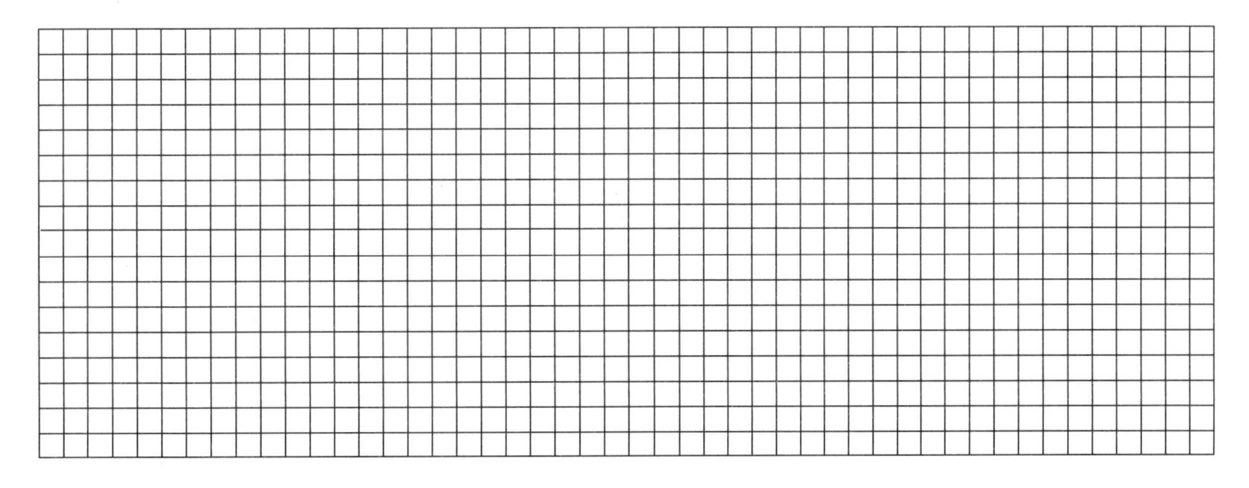

Use the Distributive Property to rewrite each equation. Find the product.

4. $12 \times 18 = n$

$n =$ ______________

5. $20 \times 23 = n$

$n =$ ______________

6. $30 \times 33 = n$

$n =$ ______________

Mixed Applications

7. A tennis player serves the ball at a speed of 95.42 miles per hour. What does the 2 digit stand for?

____

8. Every week, Jan runs 19 miles on Monday through Thursday and 17 miles on Saturday. How many miles does she run in a week?

____

Name ______________________

LESSON 6.2

Multiplying by Two-Digit Numbers

Find the product.

1. 16×12
2. 17×15
3. 18×14
4. 20×18
5. 24×16
6. 45×25
7. 80×35
8. 125×30
9. 24×46
10. 16×37
11. 43×54
12. 74×47

Mixed Applications

13. A bus company has 36 buses. Each bus can hold 54 people. How many people can the bus company serve in all?

____

14. Ralph's car gets 24 miles to a gallon of gas. Its tank holds 16 gallons of gas. How far can Ralph's car go on one tank of gas?

____

15. Seth, Brian, and Mark are comparing their heights. At 52 inches, Seth is 6 inches taller than Brian. Brian is 3 inches shorter than Mark. How tall is Mark?

____

16. Racers can choose from 4 brands of sports drinks. Each brand comes in 5 flavors. What is the total number of drink choices?

____

Name ________________________________

Estimating Products

Estimate the product by rounding each factor to its greatest place-value position.

1. 104×35 ___ × ___	2. 125×49 ___ × ___	3. 162×66 ___ × ___
4. 175×75 ___ × ___	5. 215×51 ___ × ___	6. 348×81 ___ × ___
7. 455×55 ___ × ___	8. 496×64 ___ × ___	9. 549×49 ___ × ___

10. $38 \times 562 = n$ ____________

11. $35 \times 735 = n$ ____________

12. $86 \times 860 = n$ ____________

13. $56 \times 479 = n$ ____________

14. $28 \times 358 = n$ ____________

15. $42 \times 734 = n$ ____________

Mixed Applications

16. At the garden center, there were 174 rows of flowers. Each row contained 86 flowers. Estimate the number of flowers in all.

____

17. Joel's dad sold his paintings for $750 each. He sold 26 paintings at a show. Estimate how much money in all Joel's dad made.

____

18. The Palmers have a large freezer that measures 6 feet long by 2 feet wide and 2 feet tall. What is the volume of the freezer?

____

19. Sharon spent $2 on a magazine; then she spent half of the rest of her money on shoes. She spent half of her remaining money on lunch. She has $12 left. How much did Sharon start with?

____

Name ____________________

LESSON 6.4

Multiplying by Three-Digit Numbers

Estimate to the greatest place-value position. Then find the product.

1. 126×115 $\times$ ____

2. 174×149 $\times$ ____

3. 243×176 $\times$ ____

4. 339×264 $\times$ ____

5. 456×349 $\times$ ____

6. 627×495 $\times$ ____

7. $1{,}495 \times 627$ $\times$ ____

8. $2{,}501 \times 251$ $\times$ ____

9. $2{,}328 \times 346$ $\times$ ____

10. $408 \times 562 = n$

____

____

11. $329 \times 1{,}123 = n$

____

____

12. $2{,}147 \times 415 = n$

____

____

13. $336 \times 483 = n$

____

____

14. $212 \times 3{,}678 = n$

____

____

15. $4{,}552 \times 53 = n$

____

____

Mixed Applications

16. Terri consumes an average of 2,200 calories a day. About how many calories in all will she consume in a year?

____

17. Brian earns $440 a week. If he works 48 weeks a year, how much will he earn this year?

____

Name ____________________

Multiplying to Find Perimeter and Area

Multiply to find the perimeter and area for each figure.

1.

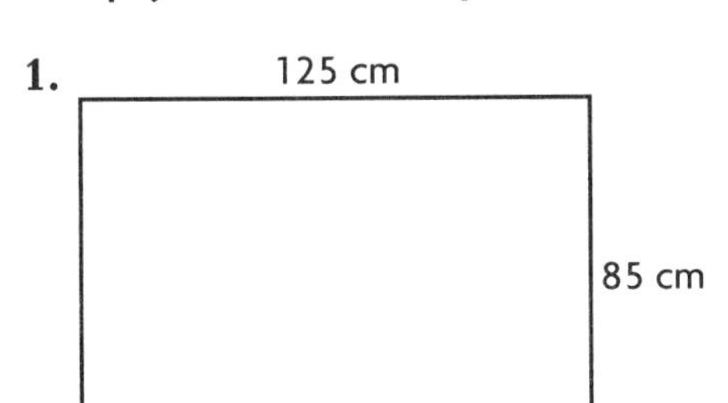

2.

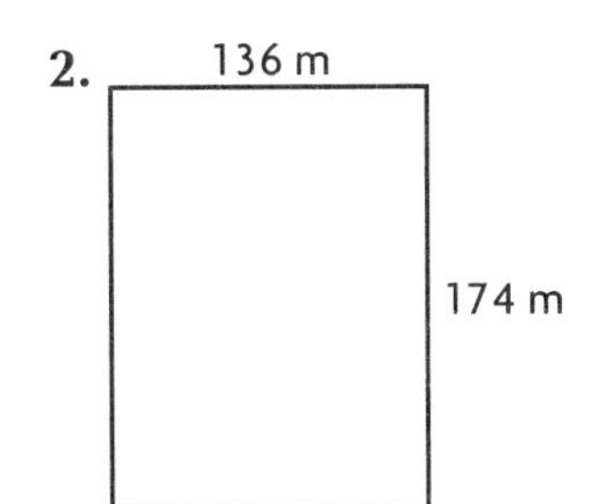

3.

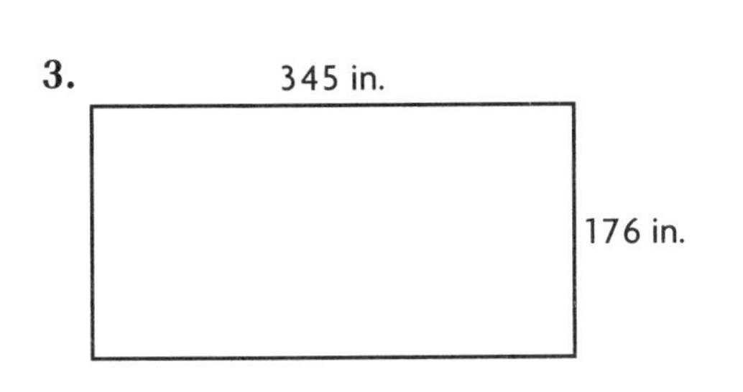

4.

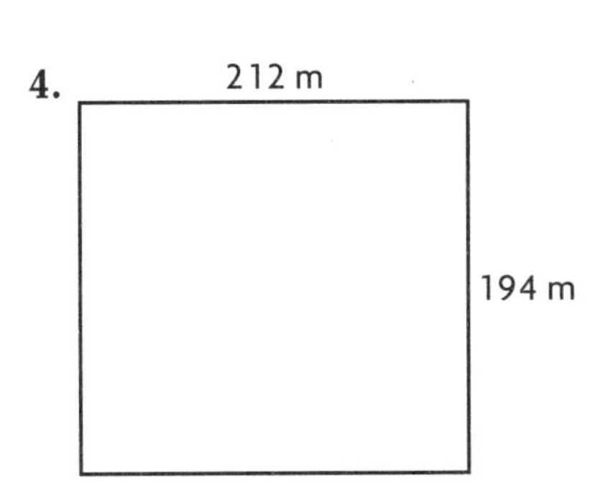

5.

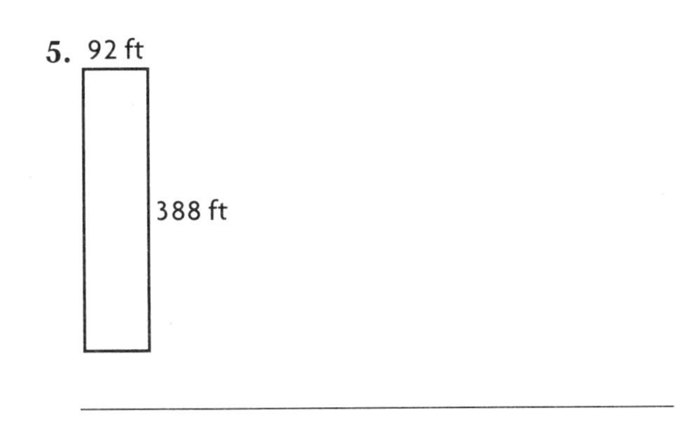

6.

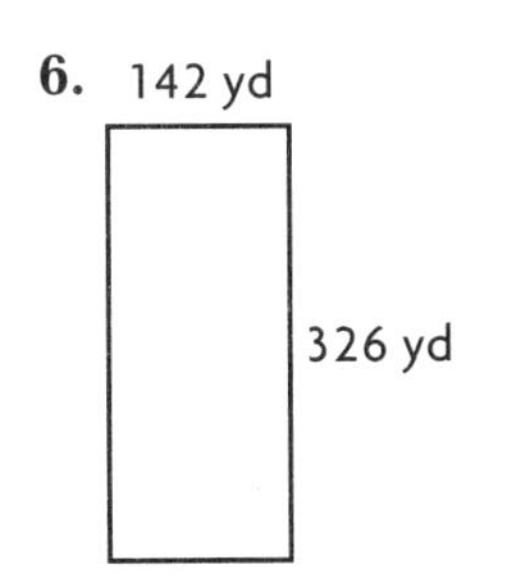

7. The school banner is 140 inches by 68 inches. What are its perimeter and area?

8. If 325 people run a 13-mile race, how many miles in all are run by the runners?

Name ______________________________

LESSON 6.5

Problem-Solving Strategy

Draw a Diagram

Draw a diagram to solve.

1. Cedric has 52 feet of wood to use to make a rectangular garden. He wants the garden to have the greatest possible area. What dimensions should it have?

____

2. Darci has 68 feet of fencing to use to make a pen for her dog. She wants her dog to have the greatest possible play area. What dimensions should the pen have?

____

3. Tamara walks 4 blocks north, 2 blocks east, and 4 blocks south. How would she walk back to where she started?

____

4. Anoki is building a fence to enclose an 18 × 24 meter rectangular area. The fence needs a post every 6 meters. How many posts does Anoki need?

____

Mixed Applications

Solve.

CHOOSE A STRATEGY

- **Draw a Diagram** • **Write a Number Sentence** • **Use a Formula** • **Make a Table** • **Guess and Check**

5. Mrs. Saturno's classroom is 23 feet × 18 feet. Mr. Downing's classroom is 22 feet × 19 feet. Whose classroom has the greater area?

____

6. It takes Ben 8 minutes to read one chapter of a book. How long will it take him to read 12 chapters?

____

7. Mt. Marcy has an elevation of 5,344 ft. The elevation of Mt. Washington is 6,288 ft; Mt. Olympus, 7,965 ft; and Mt. Katahdin, 5,268 ft. Which mountain has the highest elevation? the lowest?

____

8. Wilma has two containers. One is 10 in. × 4 in. × 8 in. The other is 9 in. × 6 in. × 6 in. Together, will they hold 700 cubic inches of jelly beans?

____

____

Name ______________________

Divisibility

Vocabulary

Fill in the blank.

1. A number is ______________ by another number if the quotient is a whole number and there is a zero remainder.

Use mental math and a calculator. Test each number for divisibility by 2, 3, 4, 5, 6, 9, and 10. List the numbers that work.

2. 54 ________
3. 144 ________
4. 420 ________
5. 864 ________
6. 990 ________
7. 1,224 ________
8. 3,600 ________
9. 6,618 ________
10. 234 ________
11. 684 ________
12. 1,827 ________
13. 2,475 ________
14. 675 ________
15. 288 ________
16. 842 ________
17. 540 ________

Mixed Applications

18. Marie made 3 dozen cookies. She needs to divide them evenly into groups greater than 4. What are all the possible equal-size groups into which she can divide the cookies?

____

19. Ted needs to divide 60 stickers into equal groups. What are all the possible equal-size groups into which he can divide the stickers?

____

____

20. A box is 7 feet long, 4 feet wide, and 3 feet high. What is the volume of the box?

____

21. I am a number between 55 and 65. I am divisible by 2, 3, 4, 5, 6, 10, and 12. What number am I?

____

Name ________________________________

LESSON 7.2

Placing the First Digit

Vocabulary

Fill in the blanks.

1. ________________________ are numbers close to the actual numbers that can be divided evenly.

Estimate the quotient.

2. $817 \div 4 \approx n$ ________

3. $462 \div 9 \approx n$ ________

4. $703 \div 7 \approx n$ ________

5. $492 \div 8 \approx n$ ________

6. $281 \div 3 \approx n$ ________

7. $539 \div 6 \approx n$ ________

8. $925 \div 3 \approx n$ ________

9. $293 \div 5 \approx n$ ________

Find the quotient.

10. $3\overline{)144}$

11. $6\overline{)174}$

12. $9\overline{)153}$

13. $7\overline{)476}$

14. $2\overline{)249}$

15. $4\overline{)855}$

16. $7\overline{)799}$

17. $8\overline{)975}$

Mixed Applications

18. Seth saved $234 during the past 6 months. Each month, he saved the same amount. How much did Seth save each month?

____

19. Rita made 150 ounces of soup. She poured the soup into 8-ounce containers. How many containers did she fill? How many ounces of soup were left over?

____

Name ______________________________

Zeros in Division

Estimate the quotient.

1. $8\overline{)330}$ 2. $6\overline{)371}$ 3. $2\overline{)813}$ 4. $9\overline{)625}$

Find the quotient.

5. $5\overline{)535}$ 6. $3\overline{)924}$ 7. $4\overline{)836}$ 8. $6\overline{)615}$

9. $2\overline{)610}$ 10. $9\overline{)960}$ 11. $7\overline{)423}$ 12. $8\overline{)647}$

13. $902 \div 9 = n$ ________
14. $409 \div 2 = n$ ________
15. $142 \div 5 = n$ ________
16. $821 \div 8 = n$ ________

Mixed Applications

17. On Saturday and Sunday, a total of 908 people attended the museum. If the same number of people came each day, how many went to the museum on Sunday?

____

18. During a 5-hour period, 510 lunches were sold in a cafeteria. If the same number of lunches were sold each hour, how many lunches were sold during the first hour?

____

19. There are 24 teams in a soccer league. If there are 15 players on each team, how many players are in the league?

____

20. I am a number between 30 and 40. I am divisible by 2, 3, 4, 6, and 9. What number am I?

____

Name ______________________

LESSON 7.4

Practicing Division

Use divisibility rules to predict if there will be a remainder.

1. $4\overline{)832}$ ________ **2.** $2\overline{)417}$ ________ **3.** $7\overline{)217}$ ________ **4.** $6\overline{)213}$ ________

Find the quotient. Check by multiplying.

5. $9\overline{)326}$ **6.** $3\overline{)235}$ **7.** $6\overline{)367}$ **8.** $4\overline{)935}$

9. $6\overline{)115}$ **10.** $9\overline{)504}$ **11.** $7\overline{)219}$ **12.** $5\overline{)621}$

13. $517 \div 2 = n$ ________ **14.** $609 \div 3 = n$ ________ **15.** $785 \div 7 = n$ ________ **16.** $431 \div 6 = n$ ________

Mixed Applications

17. On Friday and Saturday, 618 people attended a car show. If the same number of people went each day, how many people attended the car show on Saturday?

____

18. Abby rode her bike 15.73 miles. Carla rode twice as far as Abby. How many miles did Carla ride? How many miles did they ride in all?

____

19. Sue drove 364 miles in 7 hours. How many miles did she drive in 1 hour?

____

20. John paid for lunch with a $20.00 bill. He received $14.37 in change. How much money did he spend for lunch?

____

Name ______________________

Interpreting the Remainder

Solve. Explain how you interpreted the remainder.

1. A total of 134 players signed up for a soccer league. The league has 9 teams. How many players will be on most of the teams?

2. There are 230 books in the store-room. The books are stored in 7 boxes. How many books are in most of the boxes?

3. Lauren's piece of wire is 5 times longer than Larry's wire. Lauren's wire is 8 cm long. How long is Larry's wire?

4. Lee's Bakery sells muffins by the dozen. The bakery has 230 muffins prepared. Does the bakery have enough muffins to fill 20 orders?

5. Sue has 85 flowers. She put them in 7 vases with the same number of flowers in each vase except one. How many flowers are in the vase with the greatest number of flowers?

6. Jeremy had 75 feet of string. He divided it into 4 equal pieces. How long was each piece of string?

Mixed Applications

7. Each player on the soccer team has 32 raffle tickets to sell. There are 15 players on the team. How many tickets do they have to sell in all?

8. For a fund-raising project, 8 students need to sell 500 boxes of cookies. How many boxes does each student get to sell? How many boxes are left over?

Name ______________________________

LESSON 7.5

Problem-Solving Strategy

Guess and Check

Guess and check to solve.

1. Scott is 5 years old. His Aunt Mary is 4 times as old. How old will Scott be when he is half as old as his aunt?

____

2. The sum of two numbers is 42. Their product is 360. What are the two numbers?

____

3. A tunnel toll is $1.25 for cars and $2.00 for trucks. In one hour, $40.00 is collected from 23 vehicles. How many cars and trucks paid the toll?

____

4. Bob has 276 baseball cards. He keeps them in equal groups in boxes, and has started a new box with 3 cards in it. How many boxes of cards does he have? How many baseball cards are in each box?

____

Mixed Applications

Solve.

CHOOSE A STRATEGY

- **Use a Formula**
- **Write a Number Sentence**
- **Make a Table**
- **Guess and Check**

5. There were 123 cupcakes, brownies, and muffins for sale at the bake sale. There were twice as many brownies as muffins. There were 37 muffins. How many cupcakes were for sale?

____

6. Kelly rides her horse 1 hour a day 3 times a week. She rides 40 minutes a day 2 times a week. On Saturdays she rides 1 hour 15 minutes. She doesn't ride on Sundays. How much time does Kelly spend riding her horse each week?

____

7. The Scouts washed 12 cars one afternoon. They earned $6.50 for each car they washed. How much money did they earn?

____

8. The fence around Jan's backyard is 45 feet long and 30 feet wide. What is the perimeter of the fence? What is the area of the yard?

____

Name ______________________

Division Patterns to Estimate

Complete the pattern.

1. $100 \div 20 =$ ______
$1{,}000 \div 20 =$ ______
$10{,}000 \div 20 =$ ______

2. $900 \div 90 =$ ______
$9{,}000 \div 90 =$ ______
$90{,}000 \div 90 =$ ______

3. $300 \div 50 =$ ______
$3{,}000 \div 50 =$ ______
$30{,}000 \div 50 =$ ______

4. $140 \div 20 =$ ______
$1{,}400 \div 20 =$ ______
$14{,}000 \div 20 =$ ______

5. $250 \div 50 =$ ______
$2{,}500 \div 50 =$ ______
$25{,}000 \div 50 =$ ______

6. $360 \div 60 =$ ______
$3{,}600 \div 60 =$ ______
$36{,}000 \div 60 =$ ______

Find the quotient.

7. $120 \div 40 =$ ______ **8.** $240 \div 80 =$ ______ **9.** $810 \div 90 =$ ______

10. $350 \div 70 =$ ______ **11.** $480 \div 60 =$ ______ **12.** $720 \div 80 =$ ______

13. $4{,}000 \div 80 =$ ______ **14.** $2{,}000 \div 20 =$ ______ **15.** $4{,}500 \div 50 =$ ______

16. $21{,}000 \div 70 =$ ______ **17.** $63{,}000 \div 90 =$ ______ **18.** $54{,}000 \div 90 =$ ______

19. $36{,}000 \div 40 =$ ______ **20.** $12{,}000 \div 30 =$ ______ **21.** $24{,}000 \div 80 =$ ______

Mixed Applications

22. A classroom in Appleby School is 60 feet long. How long would a row of 30 classrooms be?

____

23. There are 60 boxes of pencils in the school supply closet. Each box holds 50 pencils. How many pencils are there in all?

____

24. In the month of April, Sam delivered a total of 1,200 newspapers during the 30 days. He delivered the same number of papers each day. How many papers did Sam deliver daily?

____

25. Mark bought an ice-cream sundae for himself and one for each of his 3 friends. Each sundae cost \$1.75. How much change did Mark get back from his \$10 bill?

____

Name ______________________

LESSON 8.2

Estimating Quotients

Write two pairs of compatible numbers for each. Give two possible estimates.

1. $359 \div 55 = n$ ________ ________
2. $715 \div 74 = n$ ________ ________
3. $156 \div 37 = n$ ________ ________
4. $438 \div 57 = n$ ________ ________
5. $1,893 \div 52 = n$ ________ ________
6. $3,127 \div 44 = n$ ________ ________

Estimate the quotient.

7. $18\overline{)175}$
8. $37\overline{)231}$
9. $62\overline{)375}$
10. $81\overline{)255}$
11. $53\overline{)2,681}$
12. $41\overline{)3,289}$
13. $79\overline{)4,007}$
14. $29\overline{)1,811}$
15. $34\overline{)241}$
16. $53\overline{)4,787}$
17. $47\overline{)388}$
18. $68\overline{)3,594}$

Mixed Applications

19. There are 57 boxes of paper in the school supply room. A total of 3,641 sheets of paper are in the boxes. About how many sheets of paper are in each box?

____

20. Linda sold 19 boxes of cookies. Dora sold twice as many boxes as Linda sold. Eve sold 14 more boxes than Dora did. How many boxes did Eve sell?

____

21. The Hawks soccer team scored a total of 119 goals. The team played 31 games. About how many goals did the team score each game?

____

22. Seth has 423 baseball cards. His brother Jon has 117 fewer cards. How many cards does Jon have?

____

Name ______________________

Placing the First Digit

Estimate the quotient.

1. $7{,}311 \div 89 \approx n$ ______
2. $5{,}602 \div 78 \approx n$ ______
3. $2{,}109 \div 66 \approx n$ ______
4. $6{,}521 \div 77 \approx n$ ______
5. $2{,}680 \div 26 \approx n$ ______
6. $5{,}591 \div 92 \approx n$ ______

Draw a box where the first digit in the quotient should be placed.

7. $17\overline{)1{,}527}$
8. $23\overline{)1{,}941}$
9. $34\overline{)7{,}109}$
10. $45\overline{)5{,}683}$
11. $89\overline{)9{,}266}$
12. $31\overline{)6{,}683}$
13. $24\overline{)1{,}742}$
14. $87\overline{)9{,}556}$

Find the quotient.

15. $72\overline{)8{,}136}$
16. $39\overline{)4{,}579}$
17. $27\overline{)2{,}835}$
18. $49\overline{)7{,}116}$
19. $13\overline{)3{,}926}$
20. $81\overline{)9{,}446}$
21. $35\overline{)7{,}105}$
22. $26\overline{)3{,}109}$

Mixed Applications

23. A total of 560 people will be attending a dinner. The people will sit at tables of 16. How many tables should be set up?

______

24. Alison rowed the canoe from 1:15 until 3:05. Then she rowed after dinner from 6:20 until 8:40. How long did she row in all?

______

25. Larry has 612 baseball cards. He arranges his cards in albums that hold 36 cards each. How many albums has Larry filled?

______

26. A bus can hold 48 passengers. A total of 14 buses are filled to take members of the garden club on a trip. How many people went on the trip?

______

Name ______________________

LESSON 8.4

Correcting Quotients

Write *too high, too low,* or *just right* for each estimate.

1. $34\overline{)105}$ estimate 2 ______

2. $17\overline{)89}$ estimate 5 ______

3. $42\overline{)295}$ estimate 8 ______

4. $23\overline{)119}$ estimate 5 ______

5. $26\overline{)235}$ estimate 9 ______

6. $36\overline{)291}$ estimate 9 ______

7. $91\overline{)195}$ estimate 3 ______

8. $56\overline{)327}$ estimate 4 ______

Choose the better estimate to use in the quotient. Circle *a* or *b*.

9. $23\overline{)94}$ **a.** 4 **b.** 5

10. $41\overline{)173}$ **a.** 3 **b.** 4

11. $68\overline{)512}$ **a.** 7 **b.** 8

12. $58\overline{)311}$ **a.** 5 **b.** 6

Find the quotient.

13. $76\overline{)308}$

14. $23\overline{)711}$

15. $14\overline{)296}$

16. $39\overline{)177}$

17. $46\overline{)172}$

18. $29\overline{)544}$

19. $13\overline{)98}$

20. $57\overline{)382}$

Mixed Applications

21. A total of 635 people signed up for a bus trip. Each bus can hold 48 people. Will 13 buses be enough for the trip?

______

22. The glee club sold 180 tickets for its annual luncheon. At the luncheon, people will sit at tables of 8. Will 22 tables be enough?

______

23. A garden is 14 feet long and 11 feet wide. What is the area of the garden?

______

24. The Smiths traveled 312 miles on Friday and 470 miles on Saturday. How many more miles did they travel on Saturday than on Friday?

______

Name ______________________________

Using Division

Divide. Check by multiplying.

1. $16\overline{)73}$
2. $37\overline{)85}$
3. $55\overline{)92}$
4. $79\overline{)317}$
5. $35\overline{)219}$
6. $96\overline{)742}$
7. $41\overline{)265}$
8. $27\overline{)116}$
9. $71\overline{)603}$
10. $54\overline{)449}$
11. $22\overline{)391}$
12. $67\overline{)465}$
13. $63\overline{)414}$
14. $37\overline{)218}$
15. $84\overline{)761}$
16. $52\overline{)786}$

17. $4{,}581 \div 32 = n$ ______________

18. $1{,}985 \div 23 = n$ ______________

19. $8{,}042 \div 91 = n$ ______________

20. $5{,}401 \div 25 = n$ ______________

21. $1{,}933 \div 42 = n$ ______________

22. $3{,}751 \div 55 = n$ ______________

Mixed Applications

23. The students at Walnut Street School collected 3,102 cans for a recycling center. Each student brought in 6 cans. How many students attend the school?

____

24. The Sweet Shoppe sold 2,610 ice-cream cones during the 30 days of June. It sold the same number of cones each day. How many cones were sold per day?

____

25. A box is 15 cm long, 9 cm wide, and 12 cm high. What is the volume of the box?

____

26. There are 26 teams in a baseball league. Each team has 14 players. How many players are in the league?

____

Name ______________________

Choosing the Operation

Tell what operation should be used to solve each problem. Then solve.

1. Three school bands participated in a canned food drive. The Vikings collected 782 cans. The Spartans collected 1,298 cans. The Eagles collected 917 cans. How many cans did all three schools collect?

   ____

2. A basketball league gives each team a dozen basketballs to use during the season. How many basketballs are needed if there are 26 teams in the league?

   ____

3. Carlos arranges 34 blocks in a column. The total height of the column is 238 cm. How high is each block?

   ____

4. Ricky earns $3,209 a month. Diana earns $4,022 a month. How much more money per month does Diana earn than Ricky?

   ____

5. Seth needs two pieces of wire for a project. One piece must be 18 cm long, and the other piece must be 25 cm long. How much wire does Seth need for the project?

   ____

6. Barbara has 418 stamps. Her sister Jean has 56 fewer stamps. How many stamps does Jean have?

   ____

Mixed Applications

Solve.

7. At a movie theater, Gayle bought 5 adult tickets and 3 children's tickets. An adult ticket cost $5, and a child's ticket cost $3. How much money did Gayle spend?

   ____

8. A clothing company manufactured 2,496 hats. The hats were distributed evenly among 52 stores. How many hats were sent to each store?

   ____

Name ______________________

LESSON 8.6

Problem-Solving Strategy

Write a Number Sentence

Write a number sentence to solve.

1. Joel sells handmade birdhouses. He has 18 birdhouses for sale. The price of 10 birdhouses is $12 each, and 8 others sell for $10 each. If Joel sells all 18 birdhouses, how much money will he get?

2. Carole scored 9,024 on a computer game the first time she played it. Her score for the second game was 817 points less than her first score. What was Carole's score for the second game?

3. Dan bought a stereo for $972. He will pay it off in 18 months. What will his monthly payments be?

4. Rita has 645 stamps in her collection. She keeps them in albums. Each holds 48 stamps. How many albums does Rita have?

Mixed Applications

Solve.

CHOOSE A STRATEGY

- Use a Table • Act It Out • Guess and Check • Work Backward • Write a Number Sentence

5. Zach went to the mall and spent $6 on lunch, $8 on school supplies, and $5 playing video games. When he got home, Zach had $12 left. How much money did Zach start his day with?

6. Oak School's gym holds 800 students. Can students in all three grades attend an assembly in the gym at the same time? Explain.

Student Enrollment	
Sixth grade	312
Seventh grade	237
Eighth grade	276

Name ______________________________

LESSON 9.1

Finding the Median and the Mode

Vocabulary

Write the correct letter from Column 2.

Column 1

1. the middle number in an ordered series of numbers ______
2. the number that occurs the most often in a series of numbers ______

Column 2

a. median

b. mode

Use index cards to find the median and the mode for each set of data. Write each set of numbers on the index cards.

3.

Julian's Test Scores							
Test	1	2	3	4	5	6	7
Score	86	98	98	85	87	92	89

____

4.

Students' Heights					
Name	Rose	Sally	Hank	John	Raj
Inches	57	53	55	56	57

____

5.

Baseball Cards					
Name	Sam	Jen	Tad	Phil	Li
Number	300	280	320	280	340

____

6.

Magazines Sold							
Week	1	2	3	4	5	6	7
Number	180	150	175	160	225	190	225

____

Mixed Applications

7. Sara practiced doing leg-lifts for 5 days. She did 25 leg-lifts the first day, 30 leg-lifts the next three days, and 55 leg-lifts the last day. What are the median and the mode for the number of leg-lifts Sara did?

____

8. There are 2 classes with 35 students in each. There are 3 classes with 32 students in each. What is the total number of students in the 5 classes?

____

Name ______________________________

LESSON 9.2

Finding the Mean

Vocabulary

Complete.

1. The ______________, or average, is a way to find one number that represents all the numbers in a set of data.

For Problems 2–3, use the stem-and-leaf plot.

HENRY'S TEST SCORES	
Stem	**Leaves**
7	6 7 7
8	5 6 8
9	1 2 3

2. Write the mean for Henry's test scores.

____

3. Write the median and the mode.

____

For Problems 4–5, use the table.

ASSEMBLY ATTENDANCE	
Period	**Number of Students**
First	220
Second	180
Third	160
Fourth	200
Fifth	160

4. What is the mean for the number of students that attended the assembly each period?

____

5. Write the median and the mode for the data.

____

Find the mean, median, and mode for each set of data.

6. 2, 8, 3, 8, 4

____

____

7. 30, 10, 20, 10, 10

____

____

8. 115, 110, 120, 100, 100

____

____

Mixed Applications

9. There are 5 children in the Hamilton family. Their ages are 16, 14, 12, 9, 9. What are the mean, median, and mode for their ages?

____

10. There are 256 students going on a field trip. Each bus holds 30 students. How many buses will they use?

____

Name ______________________________

LESSON 9.3

Choosing a Reasonable Scale

Vocabulary

Write the vocabulary word that best describes the part of a graph.

1. a series of numbers placed at fixed distances ____________

2. the distance between each number on the scale ____________

Choose the most reasonable interval for each set of data.

3. 25, 50, 70, 75, 100 ______

4. 2, 4, 1, 7, 5 ______

5. 5, 10, 30, 40, 20 ______

6. 15, 25, 35, 20, 40 ______

a. 25

b. 5

c. 10

d. 1

Circle the letter of the more reasonable scale for the set of data.

7.

FIFTH-GRADE SURVEY	
Favorite Color	Number of Students
Red	40
Blue	50
Green	20
Yellow	10
Other	10

a. 60, 40, 20, 0

b. 50, 40, 30, 20, 10, 0

8.

CAKE SALE	
Week	Number Sold
1	10
2	5
3	15
4	12
5	20

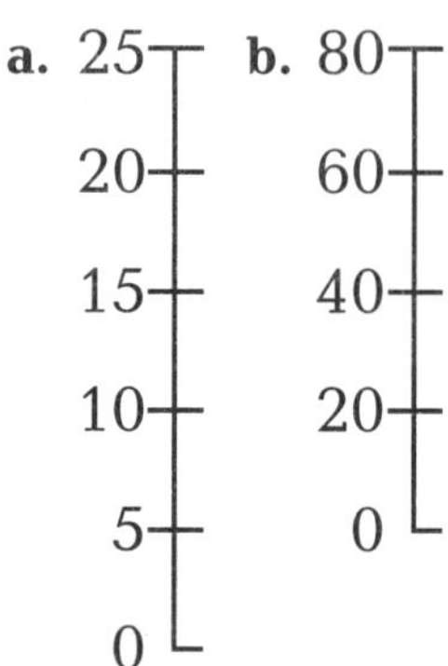

Mixed Applications

For Problems 9–10, use the table.

9. What would be an appropriate scale for these data?

____

10. How many students were surveyed?

____

SNACK SURVEY	
Favorite Snack	Number
Oatmeal cookies	18
Sandwich	20
Fruit	10

Name ______________________________

LESSON 9.4

Making Line Graphs

Vocabulary

Complete.

1. The ______________ is the difference between the greatest and least numbers in a set of data.

Make a line graph for each set of data.

2.

Books Read					
Week	1	2	3	4	5
Number of Books	15	10	20	10	5

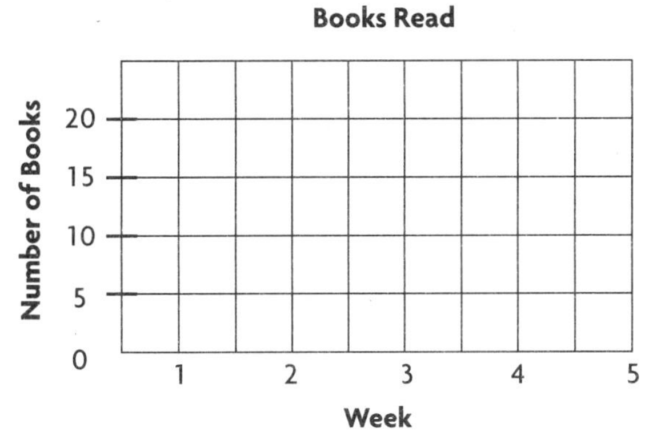

3.

Inches of Snowfall					
Month	Nov	Dec	Jan	Feb	Mar
Inches	4	12	8	6	2

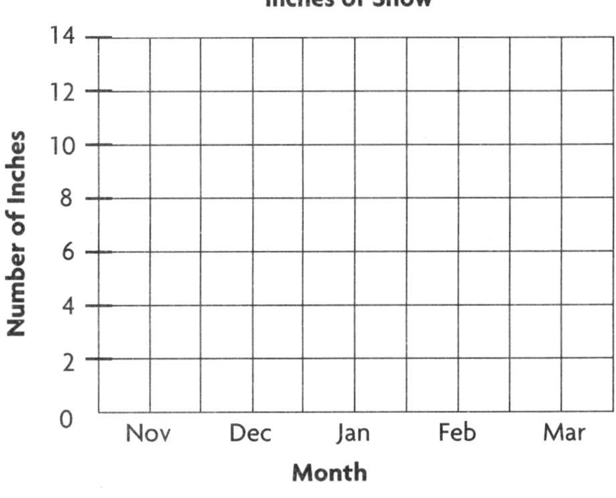

Mixed Applications

For Problems 4–5, use the table.

TIME SHEILA SPENDS PLAYING PIANO					
Day	Mon	Tue	Wed	Thu	Fri
Time	10 min	20 min	30 min	20 min	15 min

4. What would be an appropriate scale for a line graph displaying these data?

____

5. What are the mean, median, and mode for the time Sheila spent playing the piano?

____

Name ______________________________

LESSON 9.5

Choosing the Appropriate Graph

For Problems 1–4, choose a type of graph or plot. Explain your choice.

1. monthly high temperatures for a city over a 6-month period

____

____

2. heights of students in a class

____

____

3. most popular athletic shoe brand in a class

____

____

4. money spent on food each week over a 5-week period

____

____

Draw the graph or plot that best displays each set of data.

5.

Money Earned For Trip					
Week	1	2	3	4	5
Amount	\$50	\$40	\$60	\$80	\$90

6.

Favorite TV Network					
	ABC	NBC	CBS	FOX	CNN
Sixth Graders	5	10	20	20	30
Third Graders	20	15	15	30	5

Mixed Applications

For Problems 7–9, use the table.

Pets Owned in Mr. Craig's Class						
Number of Pets	0	1	2	3	4	5
Number of Students	5	7	6	8	2	1

7. What type of graph would you use to display these data? Explain.

____

8. What is the number of pets most students own?

____

9. What is the total number of students surveyed?

____

Name ______________________________

LESSON 9.5

Problem-Solving Strategy

Make a Graph

Make a graph to solve.

1.

New Mascot		
Wolf	Bear	Lion
160	140	100

Mr. Brown, the principal, surveyed students to find out which mascot they wanted. He organized the data in a table. What graph or plot should he use to display these data? Make the graph or plot.

2.

Homework Pages Assigned					
Month	Sep	Oct	Nov	Dec	Jan
Number of Pages	40	60	80	40	80

Mr. Flores kept track of the number of homework pages assigned to the class for 5 months. He recorded the data in a table. What graph or plot should he use to display these data? Make the graph or plot.

Mixed Applications

Solve.

CHOOSE A STRATEGY

- Make a Table
- Make a Graph
- Guess and Check
- Write a Number Sentence

3. Ben sold newspaper subscriptions. He sold 20 subscriptions on Monday and on Tuesday, 15 subscriptions on Wednesday and Thursday, and 30 subscriptions on Friday. What is the mean number of subscriptions Ben sold?

____

4. Samantha saved $35.50 to buy new clothes. She bought a shirt for $15.80 and a pair of pants for $12.75. How many pairs of socks priced at $1.99 a pair can she buy?

____

5. A carton has a volume of 400 cu cm. Its length is 10 cm and its width is 5 cm. What is its height?

____

6. Tracey has 4 coins in her pocket. If she has $0.46 in her pocket, what coins does she have?

____

Name ______________________________

LESSON 10.1

Reading Circle Graphs

Vocabulary

Complete.

1. A ______________ shows data as parts of a whole circle.

For Problems 2–3, use the circle graph.

Grace's 60-Minute Workout

20 min stretching before running

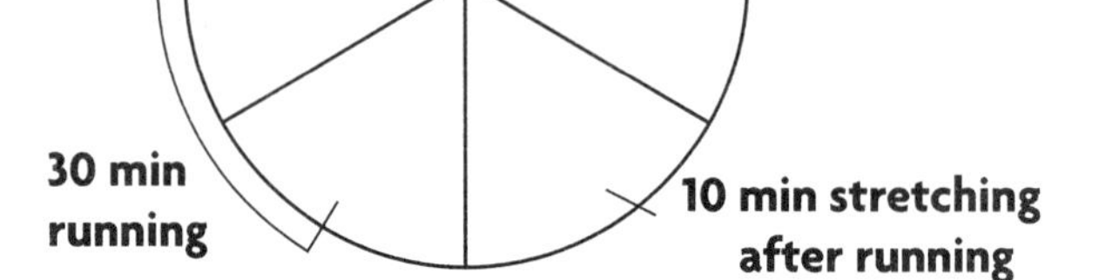

2. What does the whole circle represent?

____

3. How many minutes did Grace spend stretching?

____

Describe how the circle graph above would change for the data.

4. Grace spends 10 minutes stretching before and after running and 40 minutes running.

____

____

____

5. Grace spends 10 minutes stretching before running, 30 minutes running and 20 minutes strength training after running.

____

____

____

Mixed Applications

For Problem 6, use the circle graph.

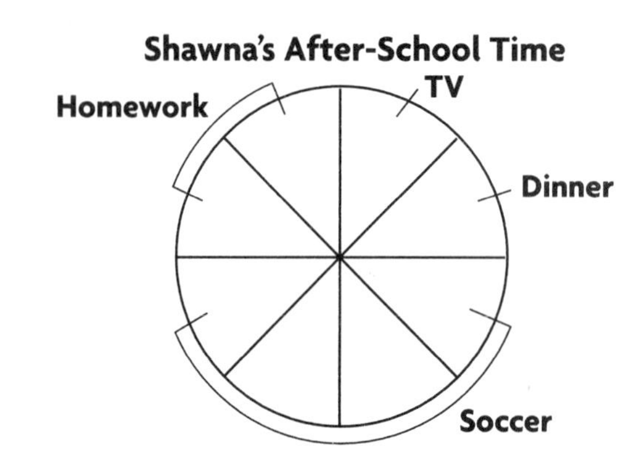

6. What fraction of the time after school does Shawna spend on soccer? on TV? on dinner? on homework?

____

Name ___________________________

LESSON 10.2

Making Circle Graphs

Use fraction-circle pieces to make a circle graph.

1.

GOALS SCORED BY HOCKEY TEAM		
Team Member	Number	Fraction of All Goals
Josh	4	$\frac{1}{4}$
Sara	4	$\frac{1}{4}$
Wiley	8	$\frac{1}{2}$

2.

BEV'S 60 CDS	
Type of CD	Fraction of All CDs
Classical	$\frac{1}{3}$
Rock	$\frac{1}{2}$
Jazz	$\frac{1}{6}$

3.

SID'S MONTHLY EXPENSES		
Items Purchased	Money Spent	Fraction of Total Spent
Snacks	$20	$\frac{1}{8}$
Hobbies	$20	$\frac{1}{8}$
Clothing	$60	$\frac{3}{8}$
Video games	$60	$\frac{3}{8}$

4.

WHAT SAM STUDIES IN 2 HOURS	
Subject	Fraction of Total Time
Science	$\frac{1}{4}$
Math	$\frac{3}{8}$
History	$\frac{1}{4}$
English	$\frac{1}{8}$

Mixed Applications

5. The yearbook staff sold 400 yearbooks at the beginning of the year and 80 yearbooks at the end of the year. If the books sold for $10 each, how much money did the yearbook staff collect?

   ____

6. Craig invited 20 people to his party. He wants to give each person a bag with 8 balloons and 5 scratch-and-sniff pencils. How many balloons and pencils does he need to buy?

   ____

7. Each yearbook has 16 pages. Paper for the yearbook is $0.07 a page. If 500 yearbooks were printed, what was the final cost for paper?

   ____

8. Samuel read 18 books in June, July, and August. He read 4 books in August. He read three times that amount in June. How many books did he read in July?

   ____

Name ________________________________

Decimals in Circle Graphs

For Problems 1–2, use the circle graph.

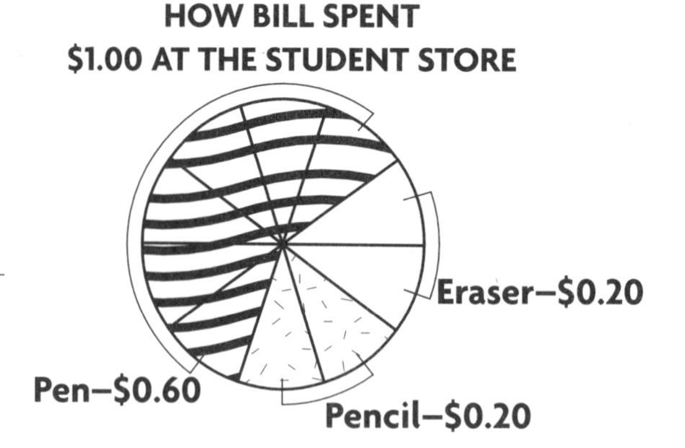

1. Bill spent $1.00 at the student store. What part of $1.00 did he spend for a pen and pencil?

____

2. What fraction represents the part of the $1.00 that Bill spent for a pen? a pencil? an eraser?

____

3. Make a circle graph for the data in the table.

SUE'S $1.00 SPENT AT THE STORE		
Item	**Amount Spent**	**Decimal Part**
Sourtarts	$0.50	0.5
Gumby Bears	$0.20	0.2
Licorice	$0.20	0.2
N and N's	$0.10	0.1

4. How would the circle graph change if Sue spent $1.00 on Sourtarts, $0.40 on Gumby Bears, $0.40 on Licorice, and $0.20 on N and N's?

____

Mixed Applications

For Problems 5–6, use the circle graph.

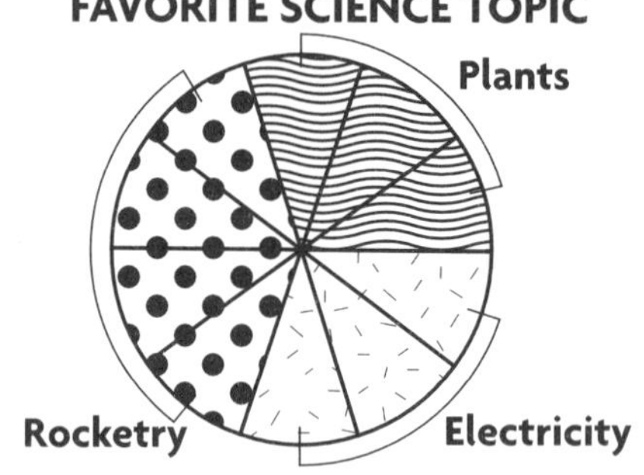

5. Write a decimal to represent the number of students out of 100 who said plants was their favorite science topic.

____

6. Write a fraction to represent the number of students out of 100 who said rocketry was their favorite science topic.

____

Analyzing Graphs

Explain why each graph does not correctly show the data.

1.

TOM'S BASEBALL CARD COLLECTION	
First Base	5 cards
Second Base	2 cards
Shortstop	1 card
Third Base	2 cards

TOM'S BASEBALL CARD COLLECTION

Number of Cards: 0, 1, 2, 3, 4, 5

First Base, Second Base, Short-stop, Third Base

Type of Card

____

2.

GROWTH OF TOM'S BASEBALL CARD COLLECTION	
June	4 cards
July	6 cards
August	10 cards
September	12 cards

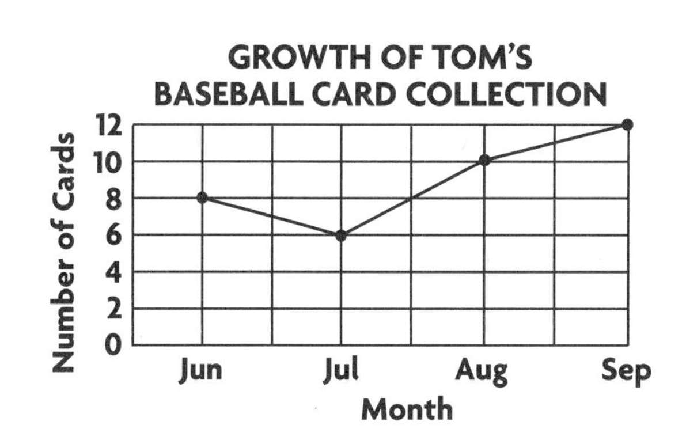

____

Mixed Applications

For Problems 3–5, use the circle graph.

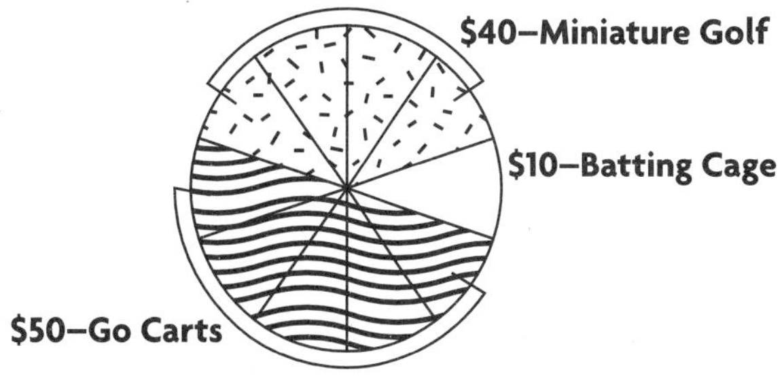

3. What decimal represents the part of the $100 earned from ticket sales for miniature golf?

____

4. What fraction represents the part of the $100 earned from ticket sales for the batting cage?

____

5. If a game of miniature golf costs $2.50, how many people played miniature golf on Family Night?

____

Name ______________________

LESSON 10.5

Comparing Graphs

Choose the best type of graph to display the data. Explain your choice.

1. amount of rainfall each month for a year in your state

2. how many newspapers were collected each week

3. the number of different types of cars on the road, such as mini-vans, sedans, wagons

4. how the Federal government spends each part of your tax dollar

5.

Monthly Sales Amounts	
Sep	$500,000
Oct	$600,000
Nov	$700,000
Dec	$900,000

6.

STUDENTS ON HONOR ROLL	
Grade	Number of Students
4th	48
5th	83
6th	64

Mixed Applications

For Problem 7, use the table.

7. What type of graph would best display these data? Explain.

Number of 100% Spelling Tests	
Week 1	12
Week 2	18
Week 3	20
Week 4	30

Name ____________________

Problem-Solving Strategy

Make a Graph

Choose the best type of graph to display the data. Then make the graph.

1. Mr. Jenning's fifth-grade class separated into three different groups to plan a party. Of the class, 0.7 worked on decorations, 0.2 worked on food, and 0.1 worked on music. What type of graph could you use to display these data?

   ____

2. Mrs. Kent's fifth-grade class went on a field trip to the zoo. For the field trip 15 students brought cameras, 8 students brought video cameras, and 7 students brought neither. What graph could you use to display these data?

   ____

Mixed Applications

Solve.

CHOOSE A STRATEGY

- Guess and Check • Act It Out • Make a Graph • Work Backward • Make a Table • Draw a Diagram

3. Sid has $0.80 in nickels and quarters. He has a total of 8 coins, and the number of nickels is 4 more than the number of quarters. How many nickels and quarters does he have?

   ____

4. The gym holds 30 rows of chairs with 28 chairs in each row. How many students can be seated in the gym?

   ____

Name ______________________

LESSON 11.1

Certain, Impossible, Likely

Vocabulary

Fill in the blanks.

1. An event is ______________ if it will always happen.

2. An event is ______________ if it will never happen.

Write *certain* or *impossible* for each event.

3. that Earth rotates

____

4. that there will be 2 Tuesdays next week

____

5. rolling a 7 on a cube numbered 1–6

____

6. that there are 13 months in a year

____

7. that the sun will rise tomorrow morning

____

8. pulling a blue counter from a bag containing six blue counters

____

Write whether each event is *likely* or *unlikely*.

9. rain falling in Seattle in winter

____

10. the cafeteria serving octopus

____

11. tossing a coin 100 times and getting heads at least 10 times

____

12. pulling the winning number out of a hat filled with 50 numbers

____

Mixed Applications

13. Patti has 6 blue bows and 1 red bow in a box. She picks 1 without looking. Is it certain or likely that she will pick a blue bow?

____

14. A magazine that costs $2.50 a month on the newsstand offers a 1-year subscription for $19.95. How much is saved by getting a subscription?

____

Name ______________________________

LESSON 11.2

Probability Experiments

Vocabulary

Fill in the blank.

1. A table of ______________________ shows results that could occur.

For Problems 2–4, use a spinner with six sections labeled 1, 1, 1, 2, 2, 3.

Spinner Experiment			
Possible outcomes			
Predicted frequency			
Actual frequency			

2. In the table at the right, list the possible outcomes.

3. Predict and record the number of times you think each outcome will occur if you spin 20 times.

____

____

4. Spin 20 times. Record the results in the table.

For Problems 5–6, use the table.

5. Why do you think you spin 1 so often?

____

6. Which number has a higher actual frequency, 2 or 3?

____

Mixed Applications

7. Chantal has a spinner divided into 4 equal sections. There are blue, red, green, and yellow sections. If she spins 10 times, what are the possible outcomes?

____

8. Marland collects cards and stores them in plastic sheets. Each sheet holds 12 cards. If Marland has 46 full sheets, how many cards does he have?

____

Name ______________________________

LESSON 11.3

Recording Outcomes in Tree Diagrams

Vocabulary

Fill in the blank.

1. A ________________ shows all the possible outcomes of an event.

Complete the tree diagram. Tell the number of choices.

Dessert	Flavors	Choices
	chocolate chip	chocolate chip cookies
2. cookies	________________	peanut butter cookies
	oatmeal	________________
	________________	chocolate cake
3. cake	vanilla	________________
	________________	strawberry cake

4. number of dessert choices: ________________

Find the number of choices by making a tree diagram.

5.

Music Choices			
Format	CD	video	tape
Type	rock	classical	jazz

number of choices: ______

6.

Book Selections		
Type	mystery	science fiction
Format	hard cover	paperback

number of choices: ______

Mixed Applications

7. Kyle wants to buy earrings for Tanya. He can buy hoop, stud, or drop earrings in gold or silver. How many choices does he have?

____

8. Joanne has $0.40 more than Cory. Together they have $9.30. How much money does each girl have?

____

____

Name ____________________

Problem-Solving Strategy

Make an Organized List

Make an organized list to solve.

1. Aber is conducting a probability experiment with a number cube and two marbles. The cube is numbered 1–6. One marble is red, the other blue. How many possible outcomes are there for this experiment? What are they?

2. Mark feeds his cat a cup of dry food and a can of wet food every day. The dry food is either chicken or fish flavored. The wet food is either tuna, salmon, or beef. List all the possible combinations of wet and dry cat food.

Mixed Applications

Solve.

CHOOSE A STRATEGY

- **Make an Organized List** • **Make a Graph** • **Work Backward** • **Act It Out** • **Guess and Check**

3. In the school election, Dave received 38 percent of the vote, Marcia received 41 percent, and Claudia received 21 percent. What type of graph would Dave use to display the data?

4. Estelle uses the numbers 3, 5, and 7 to write two-digit numbers without repeating any digits in the same number. List her numbers.

5. Martha has 6 coins that are quarters, dimes, and nickels. She has a total of $0.80. What combination of coins does she have?

6. At the movies, Jorge spent $0.95 on soda and $2.25 on popcorn. The ticket cost $4.50. If he has $2.30 left, how much money did Jorge have to begin with?

Name ______________________________

Finding Probability

Vocabulary

Fill in the blanks.

1. ______________ is the chance that an event will happen.

2. Each outcome is ________________, or has the same chance of happening.

Write a fraction for the probability of pulling each color marble from a bag of 4 red, 1 green, 2 blue, and 3 yellow marbles.

3. green ________
4. red ________
5. orange ________
6. blue ________

Write a fraction for the probability of spinning each color on a spinner with 2 red, 3 yellow, 2 green, and 1 blue sections.

7. yellow ________
8. red ________
9. yellow or blue ________
10. blue ________

Mixed Applications

11. Angie is one of 30 girls trying out for the 12 positions on the basketball team. What is the probability that Angie will make the team?

____

____

12. Of 100 tickets available for the school raffle, Tom bought 3, Jack bought 5, and Mark bought 2. What is the probability of each boy winning?

____

____

Name ______________________________

Comparing Probabilities

For Problems 1–6, you have a bag of 3 red, 5 blue, 4 yellow, and 3 green buttons. Write each probability as a fraction. Tell which outcome is more likely.

1. You pull a yellow button. ______

   You pull a red button. ______

   More likely ______

2. You pull a blue button. ______

   You pull a green button. ______

   More likely ______

3. You pull a red or yellow button.

   ______

   You pull a green or blue button.

   ______

   More likely ______

4. You pull a blue button.

   ______

   You pull a black button.

   ______

   More likely ______

5. You pull a button that isn't green. ______

   You pull a button that isn't yellow. ______

   More likely ______

6. You pull a button that isn't red.

   ______

   You pull a button that isn't blue.

   ______

   More likely ______

Mixed Applications

7. Joey had 2 pairs of red socks, 4 pairs of black socks, and 12 pairs of white socks. What is the probability that he will pull a pair of black socks from his drawer?

   ______

8. Raimondo has pizza once a week for dinner. What is the probability that he will have pizza for dinner tonight?

   ______

Name ______________________

LESSON 12.1

Multiplying Decimals and Whole Numbers

Make a model to find each product.

1. $2 \times 0.5 = n$ ______
2. $3 \times 0.4 = n$ ______
3. $2 \times 0.25 = n$ ______
4. $0.17 \times 3 = n$ ______
5. $4 \times 0.7 = n$ ______
6. $0.11 \times 4 = n$ ______
7. $3 \times 0.8 = n$ ______
8. $0.33 \times 2 = n$ ______

Phillip is buying school supplies at the student book store. For Problems 9–13, use the pictures.

9. 2 pencils, 2 erasers

____

10. 2 markers, 1 protractor

____

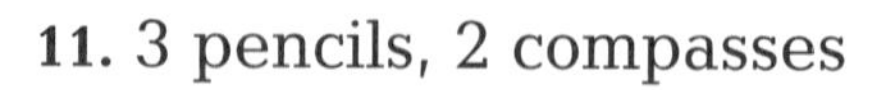

11. 3 pencils, 2 compasses

____

12. 4 markers, 2 erasers, 1 protractor

____

13. 3 compasses, 2 markers, 1 pencil

____

Mixed Applications

14. Phyllis is shopping at the student book store. Which costs more—2 markers, or 1 compass and 2 pencils?

____

15. Sam has $0.36. He has 5 coins. What are they?

____

____

Name ________________________________

LESSON 12.2

Patterns in Decimal Factors and Products

Record the decimal multiplication sentences for each.

1.

Ones	Tenths	Hundredths
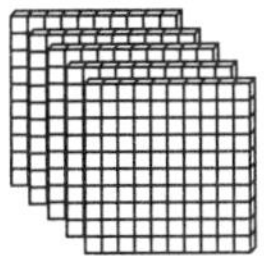		
	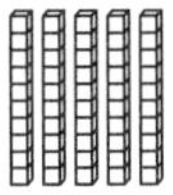	

2.

Ones	Tenths	Hundredths
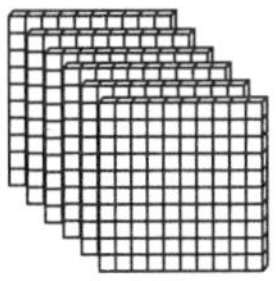		
	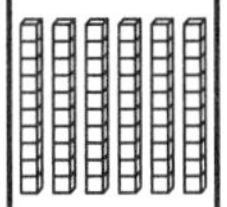	

Draw models to find each product.

3. $2 \times 2 =$ ______

$2 \times 0.2 =$ ______

$2 \times 0.02 =$ ______

4. $2 \times 5 =$ ______

$2 \times 0.5 =$ ______

$2 \times 0.05 =$ ______

Use mental math to complete the pattern.

5. $1 \times 7 = 7$

$0.1 \times 7 = 0.7$

$0.01 \times 7 =$ ____

6. $1 \times 12 = 12$

$0.1 \times 12 =$ ____

$0.01 \times 12 = 0.12$

7. $1 \times 23 = 23$

$0.1 \times 23 =$ ____

$0.01 \times 23 =$ ____

8. $1 \times 35 = 35$

$0.1 \times 35 =$ ____

$0.01 \times 35 =$ ____

Mixed Applications

9. A penny is 0.01 of a dollar. There are 50 pennies in a roll. How much is 3 rolls of pennies?

____

____

10. A dime is 0.10 of a dollar. There are 50 dimes in a roll. How much is 2 rolls of dimes?

____

____

Name ______________________

LESSON 12.3

Multiplying a Decimal by a Decimal

Complete the multiplication number sentence for each drawing.

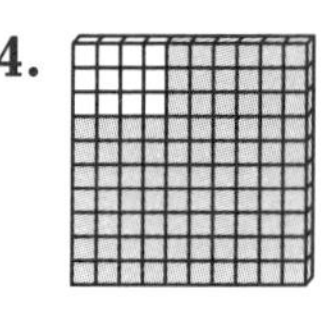

1. $0.3 \times 0.4 = n$ ______

2. $\underline{n} \times 0.7 = 0.28$ ______

3. $\underline{n} \times 0.8 = 0.16$ ______

4. $0.7 \times \underline{n} = 0.42$ ______

Multiply. Write each product.

5. $0.6 \times 0.3 =$ ______
6. $0.4 \times 0.2 =$ ______
7. $0.7 \times 0.7 =$ ______
8. $0.7 \times 0.6 =$ ______
9. $0.4 \times 0.9 =$ ______
10. $0.9 \times 0.3 =$ ______
11. $0.8 \times 0.6 =$ ______
12. $0.2 \times 0.5 =$ ______
13. $0.5 \times 0.3 =$ ______
14. $0.8 \times 0.5 =$ ______
15. $0.1 \times 0.9 =$ ______
16. $0.4 \times 0.4 =$ ______
17. $0.7 \times 0.5 =$ ______
18. $0.2 \times 0.6 =$ ______
19. $0.6 \times 0.6 =$ ______
20. $0.5 \times 0.4 =$ ______
21. $0.8 \times 0.7 =$ ______
22. $0.9 \times 0.5 =$ ______

Mixed Applications

23. Suppose you may eat 0.2 of the 0.5 cake that is left. How much can you eat?

______

24. Frank earns $4.67 each day at his job. How much does he earn in 100 days?

______

25. You order a giant cookie. Suppose you want 0.4 of the cookie to be covered with red sprinkles. If you want 0.5 of the sprinkled area also to have icing, how much of the cookie will have icing and sprinkles?

______

26. Susan bought note cards. She paid $2.00 for a set of 5 cards and $0.40 for each additional card. How much will Susan pay for 11 cards?

______

Name ______________________

Problem-Solving Strategy

Make a Model

Make a model to solve.

1. Mary has 16 ft of trim to use as a border on a blanket. She wants to make a rectangular blanket with the greatest possible area. What dimensions should the border have?

2. Sam sells school spirit stickers. He collects $1.70 in coins. If he has 4 quarters, two more dimes than quarters, and the rest in nickels, how much of each coin does he have?

3. If 1 person can sit on each side of a square table, how many people can sit at 3 tables that are pushed together end to end to form a rectangle?

4. Five friends are standing in line. Henry and Fred are next to each other and are between Lisa and John. Lisa is next to Henry. Fred and Mary are standing beside each other. Who is next to John?

Mixed Applications

Solve.

CHOOSE A STRATEGY

- **Work Backward**
- **Make a Table**
- **Find a Pattern**
- **Make a Model**

5. The product of two decimal numbers is 0.10. One number is 0.3 more than the other number. What are the two numbers?

6. Marsha has 200 flyers to hand out. If she can hand out 50 flyers in 20 minutes, how long will it take her to hand out all of the flyers?

7. Shirley buys a package of apples that weighs 3 pounds and costs $0.85 per pound. How much change will she receive from $5.00?

8. Todd earns $59.50 per week for 5 weeks. He puts $10.00 a week in his savings account. How much does he keep to spend?

Name ______________________

LESSON 12.4

Placing the Decimal Point

Vocabulary

1. What is the place value of 5 in the number 7.8645?

____

Choose the best estimate. Write *a*, *b*, or *c*.

2. $11 \times 0.3 = n$	______	a. 3	b. 30	c. 300
3. $24 \times 0.6 = n$	______	a. 1.2	b. 12	c. 120
4. $42 \times 0.9 = n$	______	a. 4	b. 40	c. 60
5. $36 \times 0.4 = n$	______	a. 0.9	b. 6	c. 15
6. $\$0.83 \times 2 = n$	______	a. $1.60	b. $16.00	c. $160.00
7. $\$0.43 \times 5 = n$	______	a. $0.20	b. $2.00	c. $4.00

Use estimation and patterns to place the decimal point in each product.

8. $2.3 \times 7 = 161$ ______

9. $2.3 \times 0.7 = 161$ ______

10. $23 \times 0.7 = 161$ ______

11. $0.23 \times 0.7 = 161$ ______

12. $23 \times 7 = 161$ ______

13. $0.23 \times 7 = 161$ ______

14. $2.3 \times 0.07 = 161$ ______

15. $23 \times 0.007 = 161$ ______

16. $0.23 \times 0.07 = 161$ ______

Mixed Applications

17. Grace earns $11.85 a week babysitting. About how much money does she earn in 12 weeks?

____

18. There is a total of 360 pieces in Hector's dinosaur puzzle. If he puts together 147 puzzle pieces, how many pieces are left to finish the puzzle?

____

Name ______________________

LESSON 12.4

More About Placing the Decimal Point

Estimate each product.

1. 6×0.21 ________
2. 7×0.41 ________
3. 0.23×95 ________
4. 64×0.4 ________
5. 0.64×9 ________
6. 32×0.04 ________
7. 0.98×46 ________
8. 82×0.2 ________

9. Complete the table.

Number Sentence	Factors	Products
$0.3 \times 0.6 = 0.18$	tenths × tenths	________
$0.3 \times$ ______ $= 0.018$	tenths × ________	________
______ × ______ $= 0.0018$	________	________
______ $\times 0.006 =$ ______	________	________

Estimate to place the decimal point. Then find the product.

10. $\$3.50 \times 2 = n$ ________
11. $2.1 \times 14 = n$ ________
12. $0.8 \times 6.7 = n$ ________
13. $1.15 \times 49 = n$ ________
14. $\$6.30 \times 7$ ________
15. 0.27×0.3 ________
16. 0.82×13 ________
17. $\$1.19 \times 97$ ________

Mixed Applications

18. Samuel buys a gallon of milk at \$1.29 and 2 dozen pretzels at \$2.19 a dozen. About how much does he spend?

____

19. John chooses a present for his father that costs \$13.95. He gives the cashier \$20.00. How much change will he receive?

____

Name ______________________

LESSON 12.5

Multiplying Mixed Decimals

Vocabulary

Complete.

1. A ______________ has a whole-number part and a decimal part in the number.

Find the product.

2. 2.84×3.4

3. 21.6×0.52

4. $\$73.49 \times 5.7$

5. 345.2×0.13

6. 401.23×3.6

7. 23.7×4.02

8. 56.1×3.8

9. $\$9.62 \times 1.5$

10. 62.4×3.05

11. 2.004×2.2

12. $1.8 \times 2.4 =$ ______

13. $1.2 \times 1.3 =$ ______

14. $2.2 \times 2.3 =$ ______

Solve. For Problems 15–16, use the table.

15. How much are 2 pairs of low top tennis shoes at Stand Up?

____

16. What is the total cost of 2 pairs of high top tennis shoes at Shoeland, including a sales tax of $0.06 on each dollar?

____

SHOE STORE PRICES

Store	Low Tops	High Tops
Shoes-R-US	$24.25	$33.50
Shoeland	$22.50	$35.75
Stand Up	$28.75	$39.75
Running Fast	$21.25	$34.50

Mixed Applications

17. Samuel bought a pen for $0.87. If he paid a sales tax of $0.06, what did he spend on the pen?

____

18. Susan was billed $1.20 for a phone call. The phone company charges $0.40 for the first minute and $0.10 for each additional minute. How long did she talk?

____

Name ______

Patterns in Decimal Division

Complete each pattern.

1. $600 \div 4 =$ ______
   $60 \div 4 =$ ______
   $6 \div 4 =$ ______

2. $100 \div 5 =$ ______
   $10 \div 5 =$ ______
   $1 \div 5 =$ ______

3. $200 \div 5 =$ ______
   $20 \div 5 =$ ______
   $2 \div 5 =$ ______

4. $300 \div 6 =$ ______
   $30 \div 6 =$ ______
   $3 \div 6 =$ ______

5. $500 \div 4 =$ ______
   $50 \div 4 =$ ______
   $5 \div 4 =$ ______

6. $1{,}200 \div 8 =$ ______
   $120 \div 8 =$ ______
   $12 \div 8 =$ ______

7. $100 \div 4 =$ ______
   $10 \div 4 =$ ______
   $1 \div 4 =$ ______

8. $1{,}400 \div 5 =$ ______
   $140 \div 5 =$ ______
   $14 \div 5 =$ ______

9. $1{,}000 \div 4 =$ ______
   $100 \div 4 =$ ______
   $10 \div 4 =$ ______

Mixed Applications

10. Phil saves 1,183 pennies. How much money does he have in dollars and cents?

    ______

11. Sid earns $60 dollars a week. He works 5 hours each week. How much does he earn per hour?

    ______

12. Theresa has 120 bows to make. She can make 6 bows in 10 minutes. How long will it take her to make all of the bows?

    ______

13. Harry needs $160 to buy a bike. He has $70. If he saves $10 each week, how many weeks will it take to save enough to buy the bike?

    ______

14. Chantel recorded in the table at the right the amount of money she spent on lunch at school for a week. On which day did she spend the most money? the least money?

    ______

Day	Amount
Monday	$2.25
Tuesday	$3.65
Wednesday	$2.05
Thursday	$1.25
Friday	$3.50

Name ______________________________

Problem-Solving Strategy

Write a Number Sentence

Write a number sentence to solve.

1. Sam and 4 classmates buy a present for their teacher. The bill for the present is $30. They share the bill equally. How much does each owe?

   ____

2. Samantha has saved $600 in her savings account. She made 10 equal deposits. How much was each deposit?

   ____

3. Each day Sam makes $8 from his dog-walking business. In 12 days, how much money will he make?

   ____

4. Cherie and Phillip worked together on a school project. Cherie spent $3.25 on supplies. Phillip spent $4.50. How much money did they spend in all?

   ____

Mixed Applications

Solve.

CHOOSE A STRATEGY

- **Make a Table**
- **Guess and Check**
- **Write a Number Sentence**

5. Henry wants to buy 6 basketball cards. At Sam's Trading Post 1 basketball card sells for $1.25. At Carl's Card Shop a set of 3 basketball cards sells for $3.60. Which store offers the better buy?

   ____

6. John wants to buy 2 new CDs. At Computers-R-Us, the CDs are 2 for $25.00. At Computerville, they are $11.95 each. What is the difference in the price for one CD?

   ____

7. Cynthia has 875 pennies and her sister Connie has 458 pennies. If the number of pennies doubles each year, how many pennies will they have in all after 2 years?

   ____

8. Sid mailed a letter. It weighed 3 ounces. The cost to mail it is 32¢ for the first ounce and 12¢ for each additional ounce. What is the cost of the letter?

   ____

Name ______________________

Decimal Division

Make a model and find the quotient.

1. 1.8 ÷ 2 = ______

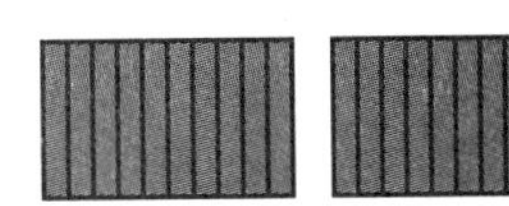

2. 2.4 ÷ 6 = ______

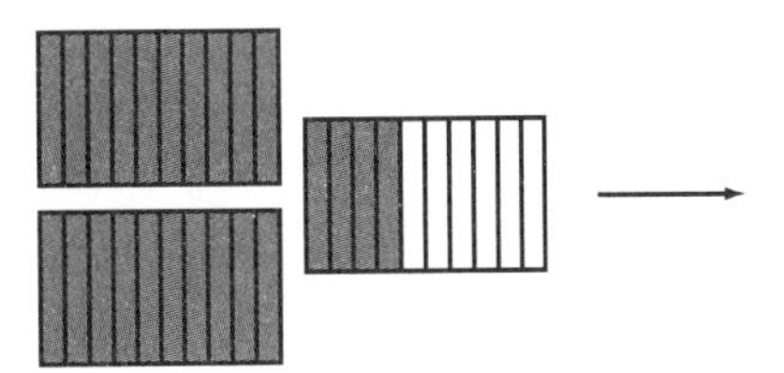

3. 0.25 ÷ 5 = ______

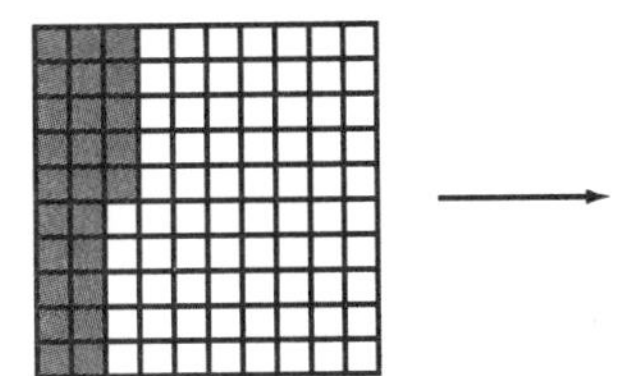

4. 0.48 ÷ 4 = ______

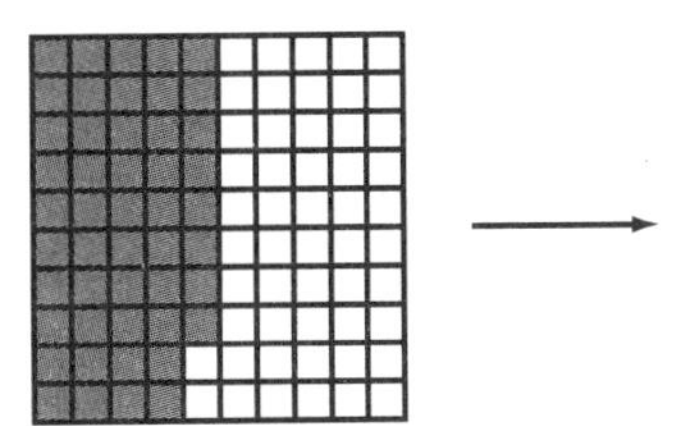

Use the model to complete the number sentence.

5.

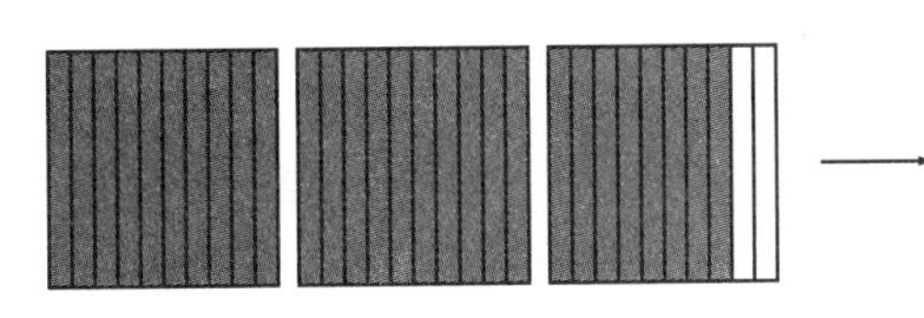

2.8 ÷ 4 = ______

Mixed Applications

6. Sam wants to decorate 4 presents. He has ribbon that is 3.2 meters long. How long will each ribbon be if he divides the ribbon evenly among the 4 presents?

____

7. Marcia is making a quilt that is 4.5 feet × 6.5 feet. What is the area of Marcia's quilt?

____

Name ______________________________

LESSON 13.3

Dividing Decimals by Whole Numbers

Find the quotient. Check by multiplying.

1. $8\overline{)4.8}$ $\times\ 8$

2. $4\overline{)2.24}$ $\times\ 4$

3. $6\overline{)4.98}$ $\times\ 6$

4. $7\overline{)47.6}$ $\times\ 7$

5. $2\overline{)6.06}$ $\times\ 2$

6. $3\overline{)2.22}$ $\times\ 3$

7. $5\overline{)6.15}$ $\times\ 5$

8. $2\overline{)16.8}$ $\times\ 2$

9. $6\overline{)8.1}$ $\times\ 6$

10. $22.4 \div 7 = \underline{?}$ ______ $\times\ 7$

11. $2.24 \div 7 = \underline{?}$ ______ $\times\ 7$

12. $0.63 \div 3 = \underline{?}$ ______ $\times\ 3$

Mixed Applications

13. At the local Farmer's Market, oranges cost $6.75 for 5 pounds. How much does 1 pound cost?

____

14. Sara is finishing her curtains. She buys 2 pieces of lace to edge the bottom. The 2 pieces cost $6.84. How much does each piece cost?

____

Name ______________________________

Placing the Decimal Point

Use estimation or patterns to place the decimal point. Then find the quotient.

1. $4\overline{)2.8}$ **2.** $5\overline{)4.5}$ **3.** $8\overline{)0.24}$ **4.** $2\overline{)0.04}$

5. $3\overline{)4.8}$ **6.** $6\overline{)10.2}$ **7.** $8\overline{)29.6}$ **8.** $9\overline{)48.6}$

9. $7\overline{)9.38}$ **10.** $4\overline{)11.04}$ **11.** $5\overline{)15.45}$ **12.** $4\overline{)21.52}$

Mixed Applications

For Problems 13–15, use the price list.

13. Sunder bought a package of hot dogs and two packages of hot dog buns. How much did he spend?

____

14. Estimate the cost of 1 pound of apples.

____

15. What is the cost of 3 packs of juice boxes?

____

Name ____________________

LESSON 13.5

Choosing the Operation

For Problems 1–4, use the table. Choose the operation and solve.

Type of Beans	Amount (in pounds)
Garbanzo beans	22.5
Black beans	64.2
Pinto beans	13.5

1. Sara buys some beans and shares each kind of bean equally with her 2 neighbors. How many pounds of garbanzo beans does Sara have after sharing with her neighbors?

____

2. How many pounds of black beans does Sara have after sharing with her neighbors?

____

3. How many pounds of pinto beans does Sara have after sharing with her neighbors?

____

4. How many pounds of beans does Sara buy in all?

____

5. Sara buys 6.4 pounds of rice each week for 4 weeks. How many pounds of rice does Sara buy?

____

Mixed Applications

6. Grant is comparing shampoos. Brand A weighs 10 ounces and sells for $3.52. Brand B weighs 8 ounces and sells for $2.16. Which shampoo costs less per ounce?

____

7. Donald bought 3 baseballs for $4.58 each, a baseball bat for $18.79, and a baseball glove for $23.52. How much did all of these items cost?

____

8. The Feldman School bought $962.50 worth of football tickets. The tickets cost $28.50 apiece. Estimate how many tickets were purchased.

____

9. Championship football tickets sold out quickly. In $1\frac{1}{2}$ hours all 6,000 tickets were sold. Estimate how many tickets were sold each minute.

____

Name ______________________________

Linear Units

Vocabulary

Write the correct letter from Column 2.

Column 1	Column 2
1. ____ millimeter(s) (mm) = 1 meter (m)	a. 100
2. ____ centimeter(s) (cm) = 1 meter (m)	b. 1
3. ____ decimeter(s) (dm) = 1 meter (m)	c. 1,000
4. ____ kilometer(s) (km) = 1,000 meters (m)	d. 10

Choose the most reasonable unit of measure. Write *mm, cm, dm, m,* or *km.*

5. length of your classroom ______________

6. thickness of a CD ______________

7. height of a portable TV ______________

8. thickness of a glass window ______________

9. length of a marathon ______________

10. length of a dollar bill ______________

Write the measurements in order from shortest to longest.

11. 2 km, 2 dm, 2 mm ______________

12. 6 m, 6 mm, 6 dm ______________

13. 72 dm, 72 m, 72 cm ______________

14. 18 m, 18 km, 18 mm ______________

Mixed Applications

15. Chantal ran in a 3,000-meter race. How many kilometers did Chantal run? ______________

16. Greg's room has a width of 3 m and a length of 4 m. Would a 150-dm border be long enough to fit around the room's perimeter? ______________

Name ____________________

Units of Mass

Vocabulary

Fill in the blanks.

1. 1 gram (g) = ______ milligrams (mg)

2. 1 kilogram (kg) = ______ grams (g)

Choose the most reasonable unit. Write *kg, g,* or *mg*.

3. a person ______

4. a T-shirt ______

5. a hamburger ______

6. a stamp ______

7. a surfboard ______

8. a scarf ______

Choose the more reasonable measurement.

9. 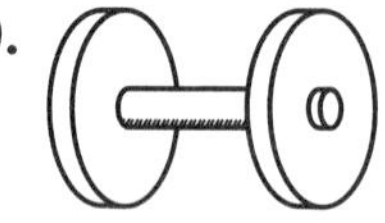

500 mg or 500 g ______

10. 

150 mg or 150 g ______

11. 

1g or 1 kg ______

12. 

15 kg or 15 g ______

13. 

250 g or 250 mg ______

14. 

60 g or 60 kg ______

Mixed Applications

15. A serving of tortilla chips contains 140 mg of salt. How many mg of salt are in 8 servings of tortilla chips?

______

16. One g of fat equals 9 calories. How many calories are from fat if a 175-calorie muffin has 5 g of fat.

______

Name ______________________________

LESSON 14.3

Units of Capacity and Volume

Vocabulary

Complete.

1. The ______________ of a container measures the amount the container will hold.

2. 1,000 ______________ = 1 liter

3. 1 ______________ = 4 metric cups

4. 1,000 liters = 1 ______________

Choose the reasonable unit. Write *mL, L,* or *kL.*

5.

____

6.
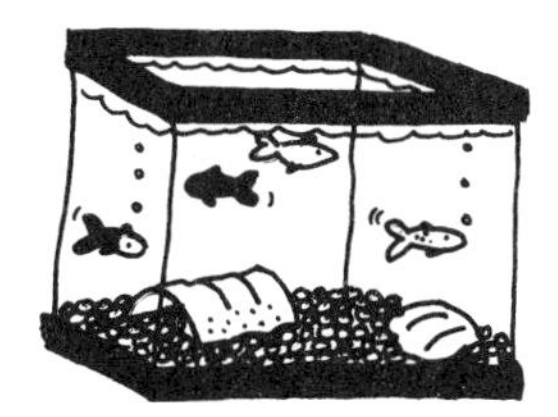

____

7.

____

Choose the more reasonable measurement.

8.

473 mL or 473 L

____

9.

4 L or 4 kL

____

10.

651 mL or 651 L

____

Mixed Applications

11. A storage shed is 14 ft long, 12 ft wide, and 10 ft high. What is the volume of the storage shed?

____

12. Aber uses 35 L of water to give her dog a bath. How many mL of water does Aber use?

____

Name ____________________

LESSON 14.4

Relating Metric Units

Vocabulary

Match the prefix in Column 2 with the base unit in Column 1.

Column 1	Column 2
1. ______ hundredths	a. deci-
2. ______ tenths	b. milli-
3. ______ thousands	c. centi-
4. ______ thousandths	d. kilo-

Choose the smaller unit of measure. Write *a* or *b*. Use the prefix to help you.

5. a. milliliter ______
   b. kiloliter

6. a. decimeter ______
   b. centimeter

7. a. gram ______
   b. kilogram

Choose the larger unit of measure. Write *a* or *b*.

8. a. liter ______
   b. kiloliter

9. a. gram ______
   b. milligram

10. a. centimeter ______
    b. millimeter

Write the equivalent measurement.

11. 900 millimeters = ______ meter

12. 0.6 meter = ______ centimeters

13. 4,500 milligrams = ______ gram

14. 3 kiloliters = ______ liters

15. 30 milliliters = ______ liter

16. 7 grams = ______ milligrams

Mixed Applications

17. There are 25 students in Emma's class. She wants to get enough juice to serve each student 300 mL. How many liters of juice does Emma need?

____

18. Sharon ran a 10-km race last weekend. Peter ran a 1,000-m race the same day. Who ran the longer race?

____

Name ______________________________

Changing Units

Write the missing unit.

1. 3 km = 3,000 ______
2. 6 L = 6,000 ______
3. 400 cm = 40 ______
4. 1.5 kg = 1,500 ______
5. 43.5 dm = 4.35 ______
6. 555 mm = 5.55 ______
7. 0.8 kL = 800 ______
8. 2 m = 200 ______

Write *multiply* or *divide*. Then write the equivalent measurement.

9. 3.5 m = ? cm ______________________________
10. 200 mL = ? L ______________________________
11. 9 L = ? mL ______________________________
12. 7 kg = ? g ______________________________
13. 6.4 g = ? mg ______________________________
14. 500 mm = ? dm ______________________________
15. 4,400 g = ? kg ______________________________
16. 800 dm = ? m ______________________________
17. 30 cm = ? dm ______________________________
18. 7.5 m = ? cm ______________________________

Mixed Applications

19. Joanne has a backpack that weighs 0.78 kg. Debby's backpack weighs 775 grams. Who has the lighter backpack?

____

20. Evan said that he walked 1,000,000 centimeters on Saturday. How far did Evan walk in meters

____

21. Tim weighs his suitcase. Would the weight be 12 g or 12 kg?

____

22. Teresa earned $178 in 6 weeks. About how much did she earn each week?

____

Name ____________________

LESSON 14.5

Problem-Solving Strategy

Draw a Diagram

Draw a diagram to solve

1. Al noticed that his cereal box has a mass of 212.5 g. A serving size is 25 g. If Al eats 1 serving of cereal every day for 2 weeks, how many boxes of cereal will he need to buy?

____

2. Gabrielle filled her 2 L water bottle. She drank 500 mL of water before her workout and 750 mL of water after her workout. How many liters of water are left in Gabrielle's bottle?

____

3. A classroom table is 1 m wide and 1.75 m long. How many 39 cm wide chairs can fit around the table?

____

4. Anna ran a marathon race, which is 42.19 km long. About how many 10,000 meter races equal the length of a marathon?

____

Mixed Applications

Solve.

CHOOSE A STRATEGY

- Guess and Check
- Make an Organized List
- Work Backward
- Write a Number Sentence
- Find a Pattern

5. Janice spent $3 for a magazine. She spent half her remaining money on school supplies. Then she spent half her remaining money on a present. She had $8 remaining. How much money did Janice begin with?

____

6. Anna walked 20 min her first day at the gym. The second day she walked 23 min. The third day she walked 26 min. If Anna continues to progress at the same rate how many minutes will she be able to walk on the twentieth day?

____

7. Egyptian surveyors used the knotted rope shown to make a right angle corner. Show how you could peg the rope to make a right triangle?

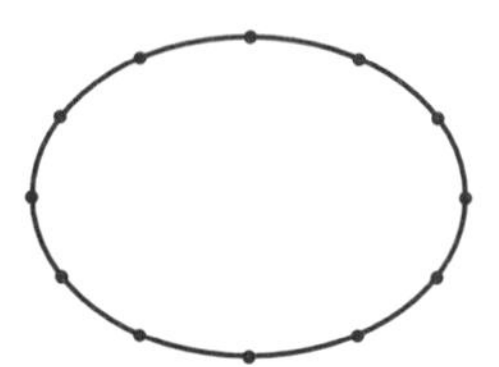

Name ________________________________

Understanding Fractions

Write the fraction shown.

1. 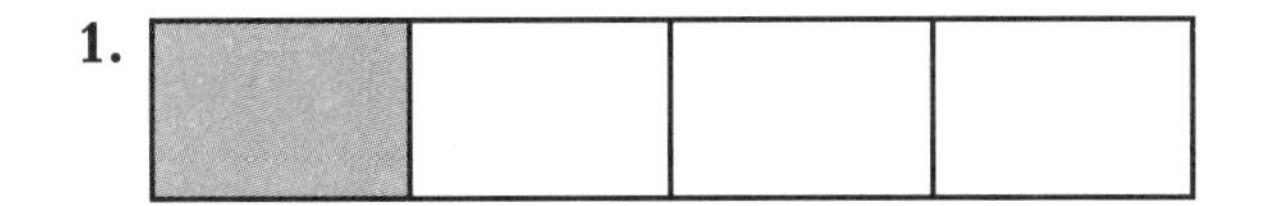

____

2. 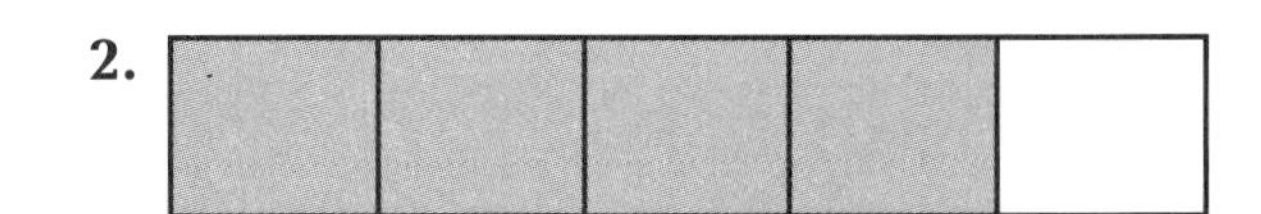

____

3. 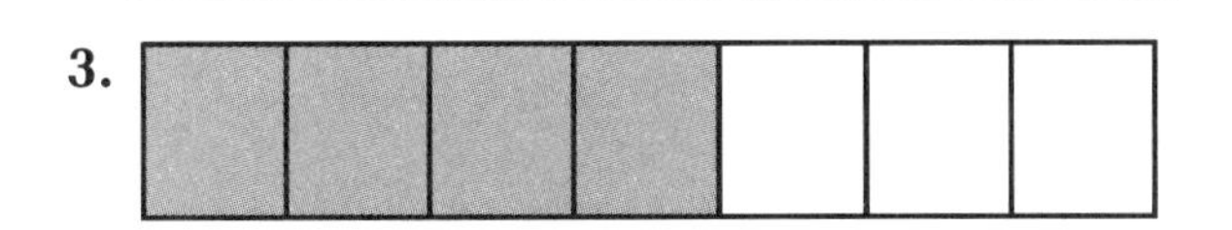

____

4. 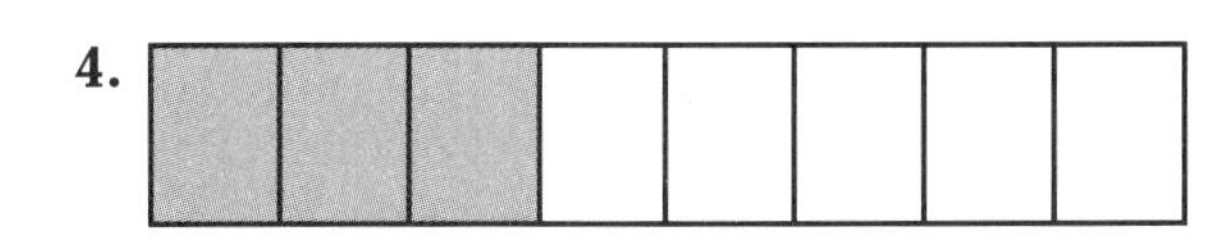

____

5. 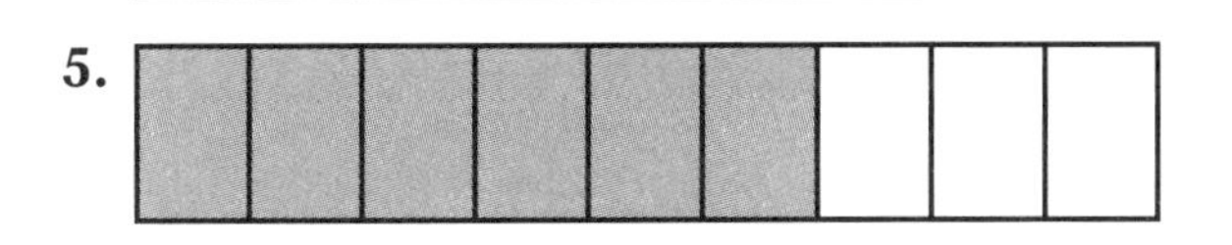

____

6. 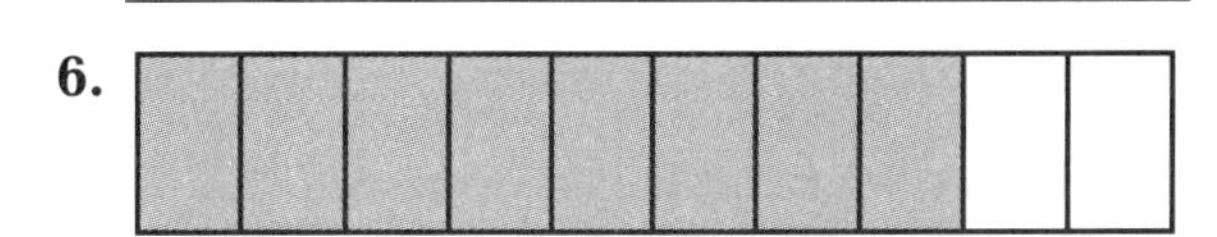

____

Shade a fraction strip to show the fraction.

7. $\frac{1}{7}$

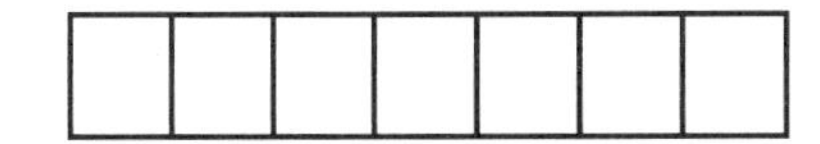

8. $\frac{4}{6}$

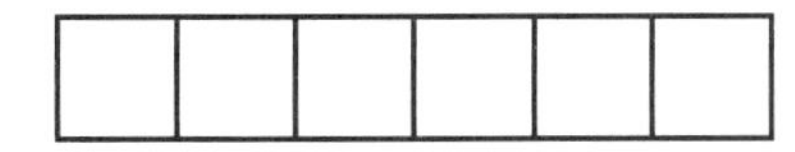

9. $\frac{5}{8}$

Draw a number line. Locate the fraction.

10. $\frac{2}{5}$

11. $\frac{7}{8}$

12. $\frac{2}{6}$

Mixed Applications

13. James has a bookcase with 6 shelves. He uses 4 of the shelves to store his books. What fraction of the shelves are not being used to store books?

____

14. Miki plans to make a sash 45 cm long and a hair band 35 cm long. She has 8 dm of ribbon. Does she have enough ribbon? Explain.

____

____

Name ____________________

LESSON 15.2

Mixed Numbers

Vocabulary

Complete.

1. A ____________________ is made up of a whole number and a fraction.

For Problems 2–5, use the figures at the right.

2. How many whole figures are shaded?

____

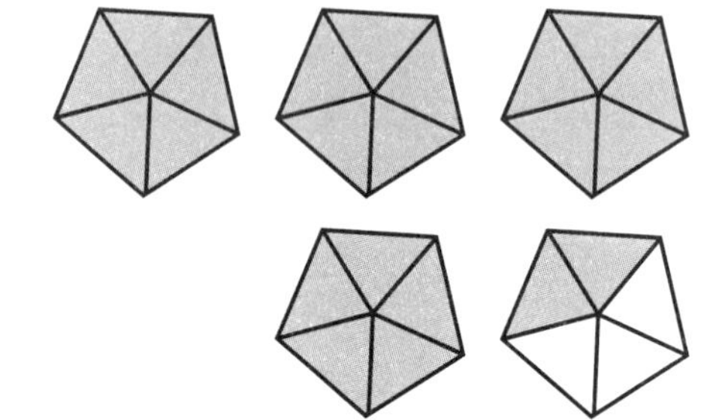

3. Into how many parts is each figure divided?

____

4. How many parts of the fifth figure are shaded?

____

5. Write a fraction and a mixed number for the figures.

____

Rename each fraction as a mixed number.

6. $\frac{22}{7}$ __________ 7. $\frac{7}{5}$ __________ 8. $\frac{19}{4}$ __________ 9. $\frac{13}{2}$ __________

Rename each mixed number as a fraction.

10. $4\frac{2}{3}$ __________ 11. $1\frac{4}{6}$ __________ 12. $3\frac{2}{5}$ __________ 13. $2\frac{2}{4}$ __________

Mixed Applications

14. Sam was asked to change $\frac{11}{3}$ to a mixed number, and $8\frac{1}{3}$ to a fraction. What answers should Sam give?

____

15. Maria takes 6 classes. In 5 of those classes, she has an A. Write a fraction to describe the fraction of classes in which she has an A.

____

Name ____________________

LESSON 15.3

Multiples and Least Common Multiples

Vocabulary

Complete.

1. The product of two or more numbers is a ____________________.
2. Multiples of one number that are also multiples of another number are called ____________________.
3. The smallest number that is a common multiple is called the ____________________, or ____________________.

Use counters to name the least common multiple for each.

4. 3 and 4 ________
5. 2 and 6 ________
6. 4 and 5 ________
7. 8 and 6 ________
8. 3 and 7 ________

Rename each pair of fractions so they have the same denominator. Use counters and LCMs from Exercises 4–8.

9. $\frac{1}{3}$ and $\frac{1}{4}$ ________
10. $\frac{1}{2}$ and $\frac{2}{6}$ ________
11. $\frac{3}{4}$ and $\frac{2}{5}$ ________
12. $\frac{3}{8}$ and $\frac{5}{6}$ ________
13. $\frac{2}{3}$ and $\frac{6}{7}$ ________
14. $\frac{5}{6}$ and $\frac{1}{4}$ ________

Mixed Applications

15. Max uses $\frac{7}{12}$ of a bag of toothpicks to make a project. Tracy uses $\frac{5}{8}$ of a bag for her project. Rename the amount of bags that Max and Tracy use so that the fractions have the same denominator.

____

16. Janet is planting a garden. She wants to put a fence around it to keep the rabbits out. If the garden is 3 meters wide and 6.5 meters long, how much fencing does she need?

____

Name ____________________

Comparing

Rename, using the least common multiple; then compare. Write <, >, or = in each ○.

1. $\frac{3}{12}$ ○ $\frac{5}{8}$
2. $\frac{2}{8}$ ○ $\frac{7}{32}$
3. $\frac{6}{8}$ ○ $\frac{3}{9}$
4. $\frac{2}{3}$ ○ $\frac{6}{9}$
5. $\frac{5}{6}$ ○ $\frac{3}{4}$
6. $\frac{3}{15}$ ○ $\frac{1}{3}$
7. $\frac{6}{22}$ ○ $\frac{3}{11}$
8. $\frac{3}{7}$ ○ $\frac{6}{21}$
9. $\frac{5}{6}$ ○ $\frac{5}{8}$
10. $\frac{3}{7}$ ○ $\frac{11}{14}$
11. $\frac{7}{12}$ ○ $\frac{3}{8}$
12. $\frac{9}{10}$ ○ $\frac{6}{7}$
13. $\frac{12}{40}$ ○ $\frac{6}{10}$
14. $\frac{4}{5}$ ○ $\frac{2}{4}$
15. $\frac{4}{7}$ ○ $\frac{1}{2}$
16. $\frac{3}{4}$ ○ $\frac{8}{9}$

Mixed Applications

17. Jacqueline is going to visit her relatives, who live 958.6 km away. On Monday, she drives 343.7 km. She drives 272.4 km on Tuesday and on Wednesday. How far must she drive on Thursday?

____

18. Every week Shelby takes a math quiz. The first week she got 17 correct. Then for the next four weeks, she got 6 more correct than on the previous week's quiz. How many questions did she get correct on her last quiz?

____

19. Doug painted $\frac{3}{5}$ of a design blue. He painted $\frac{6}{15}$ of it green. Which color is used more in his design?

____

20. If Nathan buys 3 books at $3.95 each and 2 magazines at $2.75 each, how much change will he receive if he gives the cashier $20?

____

21. Petra loves animals. She has twelve pets in all, four of which are rabbits. Write a fraction to describe the number of rabbits she has.

____

22. Flora's Flowers sells 3 roses for $13.50. The Green Thumb sells 4 roses for $15.00. Discount Flowers sells 6 roses for $23.00. Who sells roses at the lowest price?

____

Name ______________________

Ordering

Rename the fractions, using the LCM as the denominator.

1. $\frac{1}{3}, \frac{1}{4}, \frac{1}{5}$ ______

2. $\frac{5}{6}, \frac{3}{4}, \frac{5}{8}$ ______

3. $\frac{2}{5}, \frac{1}{3}, \frac{3}{5}$ ______

4. $\frac{1}{6}, \frac{1}{9}, \frac{1}{3}$ ______

5. $\frac{1}{2}, \frac{3}{8}, \frac{5}{12}$ ______

6. $\frac{3}{4}, \frac{5}{8}, \frac{2}{3}$ ______

Write in order from least to greatest.

7. $\frac{2}{5}, \frac{2}{3}, \frac{4}{15}$ ______

8. $\frac{2}{3}, \frac{3}{4}, \frac{7}{12}$ ______

9. $\frac{7}{9}, \frac{1}{2}, \frac{11}{18}$ ______

Write in order from greatest to least.

10. $\frac{5}{6}, \frac{1}{4}, \frac{5}{12}$ ______

11. $\frac{4}{5}, \frac{7}{10}, \frac{1}{2}$ ______

12. $\frac{9}{15}, \frac{2}{3}, \frac{2}{5}$ ______

Mixed Applications

13. Gilbert is making snack mix for a school field trip. He uses $\frac{1}{2}$ cup dried apricots, $\frac{3}{5}$ cup peanuts, and $\frac{4}{7}$ cup raisins. List the ingredients in order from least to greatest.

______

14. Marisa wants to buy 1 pound of walnuts and $\frac{1}{2}$ pound of almonds. She has \$12.00. If $\frac{1}{4}$ pound almonds costs \$2.25 and $\frac{1}{3}$ pound walnuts cost \$1.75, does she have enough money?

______

Name ____________________

LESSON 15.5

Problem-Solving Strategy

Draw a Diagram

Draw a diagram to solve.

1. Three students want to compare the heights of the plants they are growing in class. Bill's plant is $\frac{11}{12}$ feet tall, Karen's is $\frac{3}{4}$ feet tall, and Lois' is $\frac{5}{6}$ feet tall. List the heights from least to greatest.

    ____

2. Charlie cut a cake into 18 pieces. He decorated $\frac{2}{9}$ of the pieces blue, $\frac{1}{3}$ yellow, $\frac{2}{6}$ green, and $\frac{1}{9}$ red. How many pieces of each color are there?

    ____

    ____

3. Peter shared a bag of 16 pencils with his friends. He gave $\frac{1}{4}$ to Mark, $\frac{3}{8}$ to Sue, and the rest to Pat. How many pencils did Pat get?

    ____

4. Nancy needs $\frac{2}{3}$ yard blue ribbon, $\frac{3}{8}$ yard white fabric, $\frac{7}{8}$ yard red felt, and $\frac{3}{4}$ yard wood for a project. List her supplies from greatest to least.

    ____

Mixed Applications

Solve.

CHOOSE A STRATEGY

- **Make an Organized List**
- **Work Backward**
- **Write a Number Sentence**
- **Find a Pattern**

5. Jeff has 4 cookies left after sharing with his friends. He gave 3 to Pete, 6 to Sam, 4 to Brett, and 2 to Sally. How many cookies did he originally have?

    ____

6. On Monday Susan walked 1 mile. On Tuesday she walked 1.5 miles; on Wednesday, 2.5 miles; and on Thursday, 4 miles. If the pattern continues, how far will she walk on Friday?

    ____

7. Mr. Wong asked the 36 students in his class what type of book they liked to read. The table at the right shows their responses. What is the most popular type of book? the least popular?

    ____

Popular Book Types	
Mysteries	$\frac{2}{4}$
Biographies	$\frac{1}{18}$
Science fiction	$\frac{4}{9}$

Name ____________________

LESSON 16.1

Prime and Composite Numbers

Vocabulary

Fill in the blanks.

1. ____________________ have exactly two factors, 1 and the number itself.

2. ____________________ have more than two factors.

Use the grid below to draw all the rectangles that can be made using each number. Each square equals one unit. Record the length and the width of each rectangle.

3. 8 ______________ ______________

4. 7 ______________ ______________

5. 12 ______________ ______________

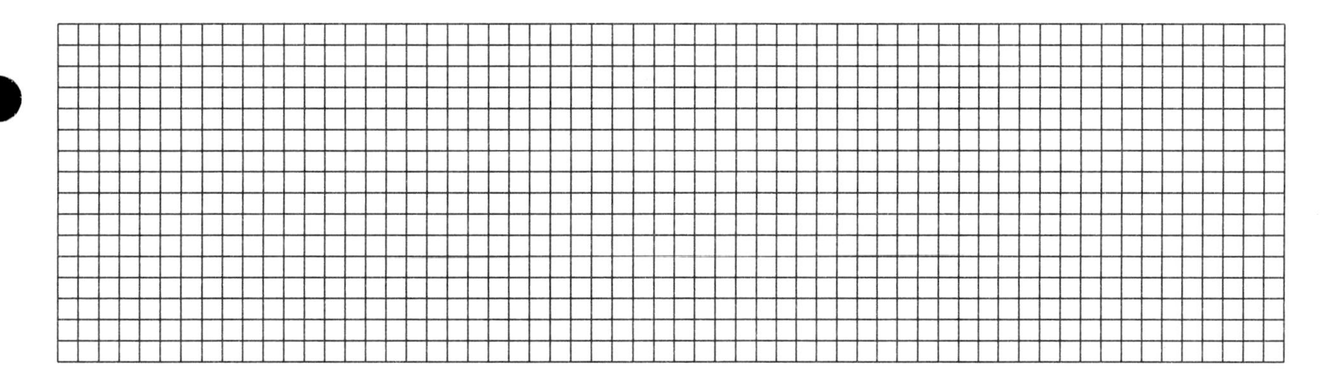

Write *prime* or *composite* for each number.

6. 11 ______________

7. 16 ______________

8. 37 ______________

Mixed Applications

9. The area of Sharon's garden is 40 sq ft. List all the possible lengths and widths of Sharon's garden.

____

____

10. Beth has $0.60 more than Suzy. Together they have $8.20. How much money does each girl have?

____

____

Name ________________________________

LESSON 16.2

Factors and Greatest Common Factors

Vocabulary

Fill in the blanks.

1. The greatest factor that two or more numbers have in common is the ______________________, or __________.

List the factors of each number.

2. 6 ____________ **3.** 20 ____________ **4.** 32 ____________

List the factors of each number. Write the greatest common factor for each pair of numbers.

5. 6 ____________
8 ____________
GCF ____________

6. 9 ____________
12 ____________
GCF ____________

7. 15 ____________
21 ____________
GCF ____________

Mixed Applications

For Problems 8–10, use the table at the right.

Cookie	Number of Cookies
Chocolate chip	24
Oatmeal	36
Raisin	16
Butterscotch	12
Lemon	28

8. The tennis team is selling packs of homemade cookies as a fund raiser. They want to pack the cookies in bags that hold the same number of one kind of cookie. What is the greatest number of cookies they can pack in each bag?

____

9. The tennis team buys empty bags in packages of 10. How many bags in all are needed to pack the cookies? How many packages of bags are needed?

____

10. The tennis team sells cookies for $0.75 per bag. If they sell 22 bags of cookies, how much money will the tennis team make?

____

Name ______________________________

Equivalent Fractions

Vocabulary

Fill in the blank.

1. Fractions that name the same amount are called ______________.

Use the number lines to name an equivalent fraction for each.

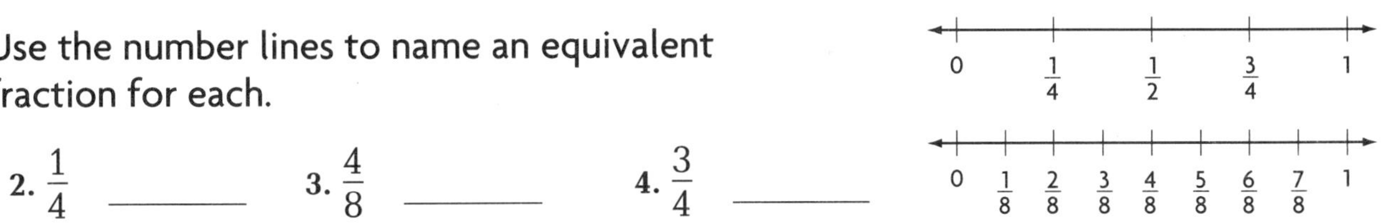

2. $\frac{1}{4}$ ________ **3.** $\frac{4}{8}$ ________ **4.** $\frac{3}{4}$ ________

Find an equivalent fraction. Use multiplication or division.

5. $\frac{2}{4}$ ________ **6.** $\frac{18}{20}$ ________ **7.** $\frac{3}{8}$ ________ **8.** $\frac{7}{21}$ ________

9. $\frac{3}{5}$ ________ **10.** $\frac{2}{15}$ ________ **11.** $\frac{8}{12}$ ________ **12.** $\frac{10}{16}$ ________

Which fraction is *not* equivalent to the given fraction? Circle *a*, *b*, or *c*.

13. $\frac{2}{3}$ **a.** $\frac{6}{9}$ **b.** $\frac{5}{6}$ **c.** $\frac{8}{12}$

14. $\frac{9}{15}$ **a.** $\frac{3}{5}$ **b.** $\frac{18}{30}$ **c.** $\frac{16}{25}$

15. $\frac{6}{8}$ **a.** $\frac{10}{12}$ **b.** $\frac{3}{4}$ **c.** $\frac{24}{32}$

16. $\frac{3}{7}$ **a.** $\frac{6}{14}$ **b.** $\frac{14}{28}$ **c.** $\frac{21}{49}$

Mixed Applications

17. René and 6 friends decide to order lasagna. Each tray of lasagna is cut into 12 pieces. How many trays of lasagna will they have to buy in order for everyone to get 3 pieces? How many pieces will be left over?

____

____

18. Andy bought a pack of 16 pencils and gave 4 pencils away to friends. Write two equivalent fractions to represent the part of the pencils that Andy gave away.

____

Name ______________________________

LESSON 16.3

Problem-Solving Strategy

Draw a Diagram

Draw a diagram to solve.

1. At a baseball game, Ben spent $\frac{4}{12}$ of his money on tickets and $\frac{1}{4}$ of his money on food. On which item did Ben spend more money? Explain how you know.

____

____

2. Juliana bought 21 drinks, 14 of which are fruit punch. Toby said that $\frac{2}{3}$ of the drinks are fruit punch. Is Toby right? Explain.

____

3. Alan bought a package containing 16 peaches. He gave 4 peaches to his uncle and 8 to his mother. Write two equivalent fractions to describe the fraction of the package of peaches that is left.

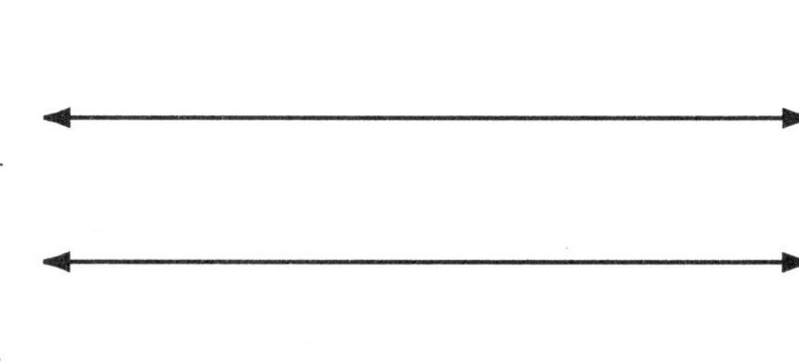

____

Mixed Applications

Solve.

CHOOSE A STRATEGY

- **Draw a Diagram**
- **Find a Pattern**
- **Make a Model**
- **Guess and Check**
- **Work Backward**

4. Julie spent half her money on lunch. Then she spent half her remaining money on a magazine. She has $4.50 left. How much money did Julie begin with?

____

____

5. The length of Alice's rectangular garden is twice its width. The perimeter of the garden is 48 feet. What are the length and width of the garden?

____

____

Name ______________________________

LESSON 16.4

Simplest Form

Is the fraction in its simplest form? Write *yes* or *no*.

1. $\frac{3}{4}$ ______ **2.** $\frac{6}{8}$ ______ **3.** $\frac{7}{21}$ ______

4. $\frac{14}{15}$ ______ **5.** $\frac{12}{15}$ ______ **6.** $\frac{7}{9}$ ______

Write each fraction in its simplest form.

7. $\frac{4}{10}$ ______ **8.** $\frac{3}{8}$ ______ **9.** $\frac{6}{12}$ ______

10. $\frac{6}{15}$ ______ **11.** $\frac{2}{3}$ ______ **12.** $\frac{4}{16}$ ______

13. $\frac{2}{8}$ ______ **14.** $\frac{8}{12}$ ______ **15.** $\frac{8}{24}$ ______

16. $\frac{3}{9}$ ______ **17.** $\frac{4}{15}$ ______ **18.** $\frac{7}{17}$ ______

Mixed Applications

19. Four students stand in line. Ken is after Jim and before Gary. Andy is after Jim and before Ken. If Ken is next to Gary, in what order are the students?

____

20. Gil bought 6 CDs on Friday. He now has 48 CDs. What fraction of his CDs did he buy on Friday? Write your answer in the simplest form.

____

21. Jean-Paul uses $\frac{1}{3}$ cup walnuts, $\frac{1}{8}$ cup chocolate chips, and $\frac{1}{2}$ cup coconut in his cookie recipe. Of these ingredients, which is used the most? Use fraction bars to explain your answer.

____

____

____

22. Mary ran $\frac{3}{4}$ mile, Lila ran $\frac{2}{3}$ mile, and Sue ran $\frac{3}{8}$ mile. Who ran the farthest? Draw a diagram to solve.

____

Name ______________________________

LESSON 16.5

More About Simplest Form

Vocabulary

Fill in the blank.

1. A fraction is in ____________________ when the greatest common factor, or GCF, of the numerator and denominator is 1.

Is the fraction in simplest form? Write *yes* or *no.* If it is not in simplest form, write it in simplest form.

2. $\frac{6}{8}$ ______ 3. $\frac{2}{3}$ ______ 4. $\frac{3}{6}$ ______

5. $\frac{7}{10}$ ______ 6. $\frac{5}{30}$ ______ 7. $\frac{3}{24}$ ______

Write each fraction in simplest form.

8. $\frac{15}{30}$ ______ 9. $\frac{4}{9}$ ______ 10. $\frac{5}{40}$ ______

11. $\frac{7}{21}$ ______ 12. $\frac{10}{24}$ ______ 13. $\frac{20}{22}$ ______

14. $\frac{3}{18}$ ______ 15. $\frac{24}{42}$ ______ 16. $\frac{9}{36}$ ______

Mixed Applications

17. Dexter brought in 7 cans of food for the local food bank. His class brought in 35 cans in all. Write in the simplest form the fraction of cans Dexter brought in.

____

18. Rita received $60 for her birthday and put $36 in her savings account. Write in the simplest form the fraction of her present that Rita saved.

____

19. It takes Gloria 8 minutes to bake a batch of cookies. How long will it take her to bake 14 batches?

____

20. Shana has 6 white rabbits. She has 3 times as many brown rabbits. How many brown rabbits does she have? How many in all?

____

Name ____________________

LESSON 17.1

Adding Like Fractions

Write an addition sentence for each drawing.

1.

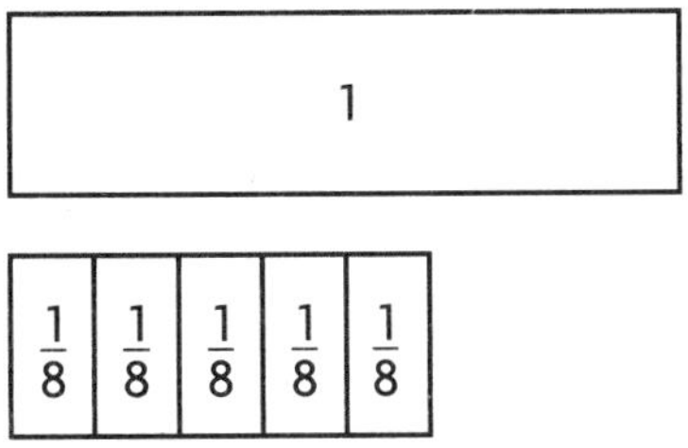

1				
$\frac{1}{8}$	$\frac{1}{8}$	$\frac{1}{8}$	$\frac{1}{8}$	$\frac{1}{8}$

____

2.

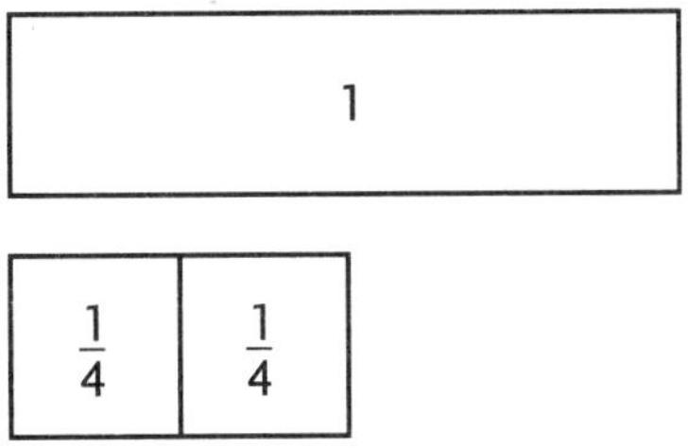

1	
$\frac{1}{4}$	$\frac{1}{4}$

____

3.

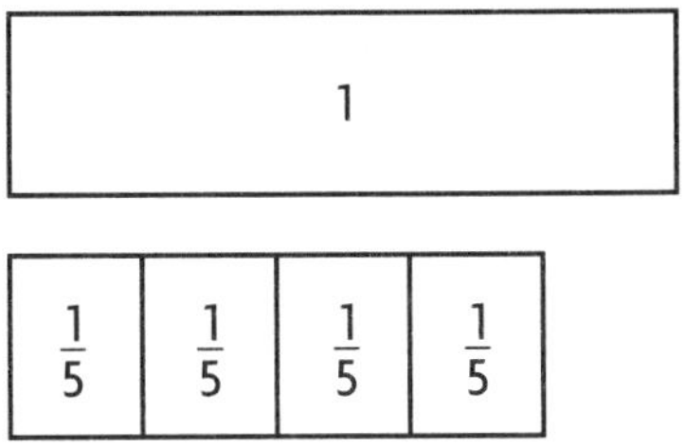

1			
$\frac{1}{5}$	$\frac{1}{5}$	$\frac{1}{5}$	$\frac{1}{5}$

____

4.

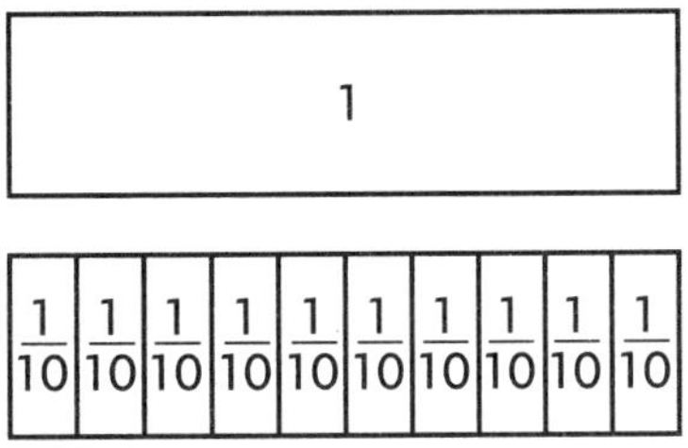

1									
$\frac{1}{10}$	$\frac{1}{10}$	$\frac{1}{10}$	$\frac{1}{10}$	$\frac{1}{10}$	$\frac{1}{10}$	$\frac{1}{10}$	$\frac{1}{10}$	$\frac{1}{10}$	$\frac{1}{10}$

____

5.

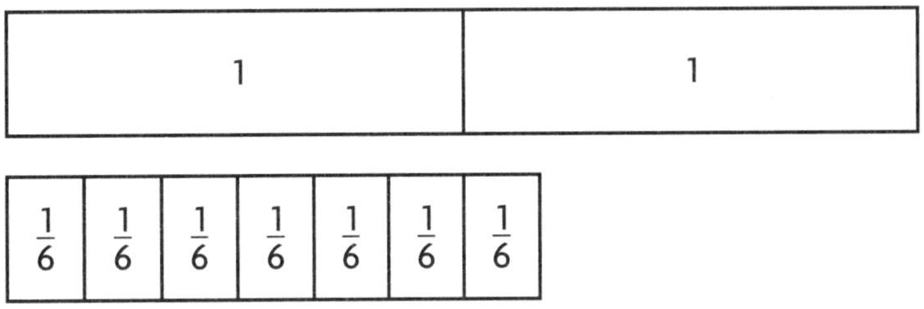

1			1			
$\frac{1}{6}$	$\frac{1}{6}$	$\frac{1}{6}$	$\frac{1}{6}$	$\frac{1}{6}$	$\frac{1}{6}$	$\frac{1}{6}$

____

Use fraction strips to find the sum. Write the answer in simplest form.

6. $\frac{2}{7} + \frac{4}{7}$ ________

7. $\frac{1}{6} + \frac{1}{6}$ ________

8. $\frac{4}{10} + \frac{7}{10}$ ________

9. $\frac{2}{3} + \frac{1}{3}$ ________

10. $\frac{3}{4} + \frac{3}{4}$ ________

11. $\frac{7}{8} + \frac{3}{8}$ ________

Mixed Applications

12. Jack bought a box of pencils. He gave $\frac{2}{7}$ to Max and $\frac{1}{7}$ to Jean. How much of the box of pencils did he give away?

____

13. Janet wants to buy 3 CDs that cost $17.99 each. How much will they cost in all?

____

Name ____________________

LESSON 17.2

Adding Unlike Fractions

Use fraction bars to find the sum.

1.

1		
$\frac{1}{3}$	$\frac{1}{3}$	$\frac{1}{6}$

______

2.

1				
$\frac{1}{4}$	$\frac{1}{4}$	$\frac{1}{8}$	$\frac{1}{8}$	$\frac{1}{8}$

______

3.

1		
$\frac{1}{3}$	$\frac{1}{3}$	$\frac{1}{4}$

______

4.

1	
$\frac{1}{2}$	$\frac{1}{5}$

______

5.

1			
$\frac{1}{12}$	$\frac{1}{12}$	$\frac{1}{12}$	$\frac{1}{3}$

______

6.

1			
$\frac{1}{10}$	$\frac{1}{10}$	$\frac{1}{10}$	$\frac{1}{5}$

______

Use fraction bars to find the sum.

7. $\frac{1}{3} + \frac{1}{6}$

______

8. $\frac{5}{8} + \frac{3}{4}$

______

9. $\frac{3}{4} + \frac{1}{6}$

______

10. $\frac{7}{10} + \frac{2}{5}$

______

11. $\frac{4}{10} + \frac{3}{5}$

______

12. $\frac{4}{5} + \frac{7}{10}$

______

Mixed Applications

13. Lewis ran $\frac{2}{3}$ of a mile on Monday and $\frac{5}{12}$ of a mile on Tuesday. How many miles did he run in all?

______

14. Alicia went to the fair with $20.00. She bought 2 drinks for $2.25 each and 1 pretzel for $1.89. T-shirts cost $5.99 each. How many T-shirts can she buy?

______

Name ____________________

Using the Least Common Denominator to Add Fractions

Vocabulary

Complete.

1. The least common multiple, or LCM, of two or more denominators is called the ____________________.

Use the LCM to name the least common denominator, or LCD, for each pair of fractions.

2. $\frac{1}{3}$ and $\frac{1}{6}$ ______

3. $\frac{1}{12}$ and $\frac{1}{4}$ ______

4. $\frac{1}{10}$ and $\frac{1}{2}$ ______

5. $\frac{1}{9}$ and $\frac{1}{3}$ ______

6. $\frac{1}{2}$ and $\frac{1}{5}$ ______

7. $\frac{1}{8}$ and $\frac{1}{2}$ ______

8. $\frac{1}{3}$ and $\frac{1}{12}$ ______

9. $\frac{1}{2}$ and $\frac{1}{4}$ ______

Use fraction strips to find the sum. Write the answer in simplest form.

10. $\frac{3}{5} + \frac{3}{10}$ ______

11. $\frac{5}{6} + \frac{7}{12}$ ______

12. $\frac{2}{3} + \frac{1}{9}$ ______

13. $\frac{1}{2} + \frac{4}{5}$ ______

14. $\frac{1}{6} + \frac{2}{3}$ ______

15. $\frac{9}{12} + \frac{1}{3}$ ______

16. $\frac{3}{4} + \frac{5}{8}$ ______

17. $\frac{2}{6} + \frac{2}{3}$ ______

Mixed Applications

18. Clara's family has 12 pets. There are 3 rabbits, 4 cats, and 5 dogs. What fraction of the pets are cats?

____

19. Roberto bought $\frac{1}{9}$ of a set of trading cards last week. He bought $\frac{2}{3}$ of the set this week. What fraction of a set does Roberto have in all?

____

Name ______________________________

LESSON 17.4

Adding Three Fractions

Use the LCM to name the least common denominator, or LCD, for each group of fractions.

1. $\frac{1}{3}, \frac{1}{4},$ and $\frac{1}{2}$ ________

2. $\frac{1}{4}, \frac{1}{6},$ and $\frac{1}{12}$ ________

3. $\frac{1}{10}, \frac{1}{2},$ and $\frac{1}{5}$ ________

4. $\frac{1}{2}, \frac{1}{6},$ and $\frac{1}{3}$ ________

5. $\frac{1}{6}, \frac{1}{3},$ and $\frac{1}{4}$ ________

6. $\frac{1}{12}, \frac{1}{6},$ and $\frac{1}{4}$ ________

Use the fraction strips to find the sum. Write the answer in simplest form.

7. $\frac{2}{3} + \frac{1}{6} + \frac{2}{6}$ ________

8. $\frac{3}{5} + \frac{7}{10} + \frac{1}{2}$ ________

9. $\frac{1}{9} + \frac{2}{3} + \frac{5}{9}$ ________

10. $\frac{1}{12} + \frac{2}{3} + \frac{1}{4}$ ________

11. $\frac{1}{3} + \frac{1}{4} + \frac{2}{6}$ ________

12. $\frac{1}{3} + \frac{3}{4} + \frac{1}{6}$ ________

13. $\frac{5}{8} + \frac{1}{2} + \frac{7}{8}$ ________

14. $\frac{4}{6} + \frac{1}{6} + \frac{3}{12}$ ________

15. $\frac{2}{4} + \frac{7}{8} + \frac{1}{4}$ ________

16. $\frac{3}{4} + \frac{2}{3} + \frac{1}{6}$ ________

17. $\frac{1}{2} + \frac{1}{4} + \frac{3}{6}$ ________

18. $\frac{1}{10} + \frac{1}{5} + \frac{1}{2}$ ________

Mixed Applications

19. An auditorium has 3 sections of seats. Each section has 15 rows with 12 seats in each row. How many people can be seated in the auditorium?

____

20. Pete sees two equal-size punch bowls. One bowl is $\frac{3}{10}$ full, and the other is $\frac{2}{5}$ full. If he pours the punch from one bowl into the other, how full is the bowl with the combined punch?

____

Name ______________________________

Problem-Solving Strategy

Make a Model

Make a model to solve.

1. Samantha bought 3 packets of stickers. Each packet contains 100 stickers. If she divides all of the stickers evenly among 6 friends and herself, how many stickers are left over?

____

2. One day, $\frac{1}{8}$ of the patients brought to a veterinary hospital were rabbits, $\frac{1}{2}$ were cats, and $\frac{1}{4}$ were dogs. What part of all the patients were rabbits, cats, and dogs that day?

____

3. James uses $\frac{5}{6}$ meter of butcher paper to make one sign. How many meters of paper will he need to make 3 signs?

____

4. Brent decorated $\frac{3}{8}$ of his sugar cookies with blue frosting, $\frac{1}{4}$ with yellow frosting, and $\frac{3}{8}$ with purple frosting. What part of the cookies were frosted?

____

Mixed Applications

Solve.

CHOOSE A STRATEGY

- **Make an Organized List** • **Work Backward** • **Act It Out** • **Use a Table** • **Make a Model**

5. During the week, Carrie spent $3.50 for a book. The next day her father gave her $1.25. Then she went to a movie, which cost $7.50. If she now has $10.25, how much money did she have at the beginning of the week?

____

6. A pizza parlor has a special offer of a minipizza with one topping. Customers can choose thin or thick crust, and they have 4 choices of toppings: pepperoni, sausage, extra cheese, or olives. How many choices do customers have?

____

Name ________________________________

LESSON 18.1

Subtracting Like Fractions

Use fraction strips to find the difference.

1. $\frac{5}{6} - \frac{4}{6} = n$ ________

$\frac{1}{6}$	$\frac{1}{6}$	$\frac{1}{6}$	$\frac{1}{6}$	$\frac{1}{6}$

2. $\frac{6}{12} - \frac{1}{12} = n$ ________

$\frac{1}{12}$	$\frac{1}{12}$	$\frac{1}{12}$	$\frac{1}{12}$	$\frac{1}{12}$	$\frac{1}{12}$

3. $\frac{4}{5} - \frac{2}{5} = n$ ________

$\frac{1}{5}$	$\frac{1}{5}$	$\frac{1}{5}$	$\frac{1}{5}$

4. $\frac{7}{8} - \frac{2}{8} = n$ ________

$\frac{1}{8}$	$\frac{1}{8}$	$\frac{1}{8}$	$\frac{1}{8}$	$\frac{1}{8}$	$\frac{1}{8}$	$\frac{1}{8}$

5. $\frac{7}{9} - \frac{2}{9} = n$ ________

$\frac{1}{9}$	$\frac{1}{9}$	$\frac{1}{9}$	$\frac{1}{9}$	$\frac{1}{9}$	$\frac{1}{9}$	$\frac{1}{9}$

6. $\frac{8}{10} - \frac{5}{10} = n$ ________

$\frac{1}{10}$	$\frac{1}{10}$	$\frac{1}{10}$	$\frac{1}{10}$	$\frac{1}{10}$	$\frac{1}{10}$	$\frac{1}{10}$	$\frac{1}{10}$

Use fraction strips to find the difference. Write the answer in simplest form.

7. $\frac{6}{8} - \frac{4}{8} = n$ ________

$\frac{1}{8}$	$\frac{1}{8}$	$\frac{1}{8}$	$\frac{1}{8}$	$\frac{1}{8}$	$\frac{1}{8}$

8. $\frac{5}{6} - \frac{2}{6} = n$ ________

$\frac{1}{6}$	$\frac{1}{6}$	$\frac{1}{6}$	$\frac{1}{6}$	$\frac{1}{6}$

9. $\frac{9}{10} - \frac{4}{10} = n$ ________

$\frac{1}{10}$	$\frac{1}{10}$	$\frac{1}{10}$	$\frac{1}{10}$	$\frac{1}{10}$	$\frac{1}{10}$	$\frac{1}{10}$	$\frac{1}{10}$	$\frac{1}{10}$

10. $\frac{11}{12} - \frac{7}{12} = n$ ________

$\frac{1}{12}$	$\frac{1}{12}$	$\frac{1}{12}$	$\frac{1}{12}$	$\frac{1}{12}$	$\frac{1}{12}$	$\frac{1}{12}$	$\frac{1}{12}$	$\frac{1}{12}$	$\frac{1}{12}$	$\frac{1}{12}$

11. $\frac{2}{4} - \frac{1}{4} = n$ ________

$\frac{1}{4}$	$\frac{1}{4}$

12. $\frac{8}{9} - \frac{5}{9} = n$ ________

$\frac{1}{9}$	$\frac{1}{9}$	$\frac{1}{9}$	$\frac{1}{9}$	$\frac{1}{9}$	$\frac{1}{9}$	$\frac{1}{9}$	$\frac{1}{9}$

Mixed Applications

13. Phil biked $\frac{2}{8}$ mile on Monday and $\frac{5}{8}$ mile on Tuesday. On which day did he bike farther? How much farther did he bike?

____

14. When Terry arrived at the party, $\frac{9}{12}$ of the cake was left. She ate $\frac{2}{12}$ of the cake. How much was left after Terry ate?

____

Name ________________

Subtracting Unlike Fractions

Use fraction bars to find the difference.

1. $\frac{1}{2}$; $\frac{1}{12}$ $\frac{1}{12}$ $\frac{1}{12}$?

2. $\frac{1}{3}$; $\frac{1}{9}$ $\frac{1}{9}$?

3. $\frac{1}{4}$ $\frac{1}{4}$ $\frac{1}{4}$; $\frac{1}{8}$ $\frac{1}{8}$?

4. $\frac{1}{3}$ $\frac{1}{3}$; $\frac{1}{12}$ $\frac{1}{12}$ $\frac{1}{12}$ $\frac{1}{12}$ $\frac{1}{12}$?

5. $\frac{1}{10}$ $\frac{1}{10}$ $\frac{1}{10}$ $\frac{1}{10}$ $\frac{1}{10}$ $\frac{1}{10}$ $\frac{1}{10}$; $\frac{1}{5}$ $\frac{1}{5}$ $\frac{1}{5}$?

6. $\frac{1}{12}$ $\frac{1}{12}$ $\frac{1}{12}$ $\frac{1}{12}$ $\frac{1}{12}$ $\frac{1}{12}$ $\frac{1}{12}$ $\frac{1}{12}$ $\frac{1}{12}$ $\frac{1}{12}$ $\frac{1}{12}$; $\frac{1}{4}$ $\frac{1}{4}$ $\frac{1}{4}$?

7. $\frac{4}{5} - \frac{3}{10} = n$

8. $\frac{4}{6} - \frac{5}{12} = n$

9. $\frac{5}{6} - \frac{5}{12} = n$

10. $\frac{1}{2} - \frac{4}{10} = n$

11. $\frac{6}{8} - \frac{1}{2} = n$

12. $\frac{2}{3} - \frac{3}{6} = n$

13. $\frac{1}{2} - \frac{1}{8} = n$

14. $\frac{9}{12} - \frac{2}{3} = n$

15. $\frac{4}{6} - \frac{1}{12} = n$

16. $\frac{7}{8} - \frac{1}{4} = n$

17. $\frac{11}{12} - \frac{1}{3} = n$

18. $\frac{4}{6} - \frac{1}{2} = n$

Mixed Applications

19. Joyce went to Camp Mattatuch for $\frac{3}{4}$ of a month. She spent $\frac{2}{8}$ of the time swimming, $\frac{1}{4}$ of the time doing crafts, and the rest hiking. What fraction of the month did she hike?

20. On a recent bicycle trip, Karen rode 75 miles in 6 days. If she rode the same number of miles each day, how many miles per day did she ride?

Name ____________________

LESSON 18.3

Using the Least Common Denominator to Subtract Fractions

Name the least common denominator, or LCD, for each pair of fractions.

1. $\frac{1}{5}$ and $\frac{1}{10}$ ________
2. $\frac{1}{3}$ and $\frac{1}{6}$ ________
3. $\frac{1}{6}$ and $\frac{1}{4}$ ________
4. $\frac{1}{2}$ and $\frac{1}{8}$ ________
5. $\frac{1}{2}$ and $\frac{1}{6}$ ________
6. $\frac{1}{5}$ and $\frac{1}{4}$ ________

Use fraction strips to find the difference. Write the answer in simplest form.

7. $\frac{5}{8} - \frac{1}{2} =$ ________
8. $\frac{2}{3} - \frac{5}{9} =$ ________
9. $\frac{5}{6} - \frac{1}{3} =$ ________
10. $\frac{9}{12} - \frac{2}{3} =$ ________
11. $\frac{5}{8} - \frac{1}{4} =$ ________
12. $\frac{4}{5} - \frac{1}{10} =$ ________
13. $\frac{5}{12} - \frac{1}{4} =$ ________
14. $\frac{4}{5} - \frac{2}{10} =$ ________
15. $\frac{4}{6} - \frac{1}{3} =$ ________

Mixed Applications

16. Tony spent $\frac{3}{4}$ of the afternoon studying for the math test. Julie spent $\frac{2}{3}$ of the afternoon studying for the test. How much longer did Tony spend studying than Judy?

____

17. Victoria was saving for a bike. She earned $\frac{3}{5}$ of the money working for her neighbors and $\frac{3}{10}$ working for her parents. The rest she got for her birthday. What fraction of the money did she get for her birthday?

____

18. Mr. Rafael bought 525 stars for his class. He wants to give each of his 35 students an equal number of stars. How many stars should each student get?

____

19. Joe drank $\frac{5}{8}$ of a glass of milk. Later in the day, he drank another $\frac{3}{4}$ glass of milk. How many glasses of milk did he drink?

____

Name ______________________

LESSON 18.4

Subtracting Fractions Using a Ruler

Use the ruler to find the difference.

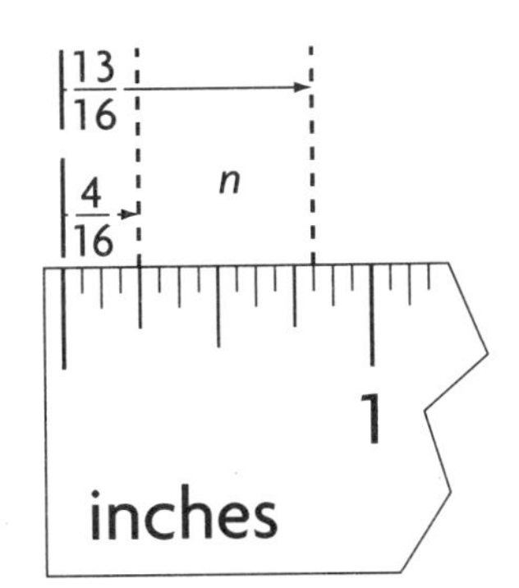

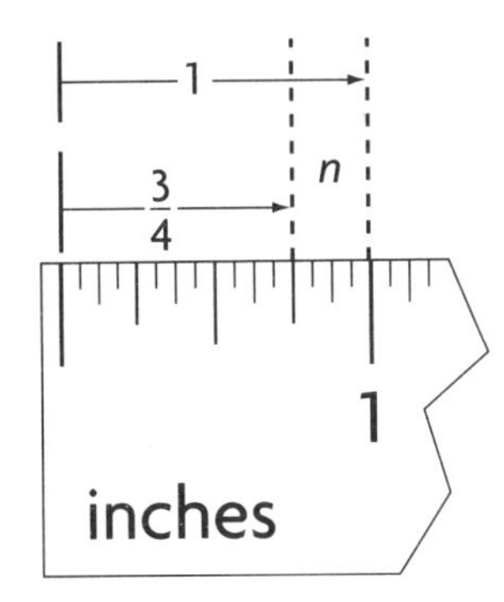

1. $\frac{13}{16}$ in. $-$ $\frac{4}{16}$ in. $= n$ ______

2. $\frac{5}{8}$ in. $-$ $\frac{2}{8}$ in. $= n$ ______

3. 1 in. $-$ $\frac{3}{4}$ in. $= n$ ______

4. $\frac{3}{8}$ in. $-$ $\frac{1}{4}$ in. $= n$ ______

5. $\frac{13}{16}$ in. $-$ $\frac{5}{8}$ in. $= n$ ______

6. $\frac{7}{8}$ in. $-$ $\frac{3}{4}$ in. $= n$ ______

7. $\frac{9}{16}$ in. $-$ $\frac{3}{8}$ in. $= n$ ______

8. 1 in. $-$ $\frac{5}{8}$ in. $= n$ ______

9. $\frac{1}{2}$ in. $-$ $\frac{1}{4}$ in. $= n$ ______

Mixed Applications

For Problems 10–11, use a ruler to solve.

10. Henry collects miniature animals. His longest animal is $\frac{15}{16}$ inch long. His shortest animal is $\frac{3}{8}$ inch long. What is the difference in length between the two animals?

______

11. Jake jumped $\frac{7}{8}$ inch farther than Ted. Phil jumped $\frac{3}{4}$ inch farther than Ted. How much farther did Jake jump than Phil?

______

Name ______________________________

LESSON 18.4

Problem-Solving Strategy

Work Backward

Work backward to solve.

1. Jerry's kitten is 19 cm tall and is 6 months old. The kitten grew 2 cm between the ages of 5 months and 6 months. It grew 3 cm between the ages of 4 months and 5 months. How tall was Jerry's kitten when it was 4 months old?

____

2. Denise went shopping at the mall. She spent $11.35 on a new T-shirt and $2.25 for hair ribbons. Lunch cost $4.50, and a drink cost $1.25. She came home with $10.65. How much money did Denise have before she went to the mall?

____

3. Kirk grew a crystal in science class. On Monday it was $\frac{13}{16}$ inch tall. It had grown $\frac{1}{4}$ inch between Friday and Monday. It had grown $\frac{1}{2}$ inch between Tuesday and Friday. How tall was Kirk's crystal on Tuesday?

____

4. Terry planted a gladiolus bulb. On Wednesday it was $\frac{7}{8}$ inch tall. It had grown $\frac{1}{4}$ inch between Tuesday and Wednesday. It had grown $\frac{3}{8}$ inch between Monday and Tuesday. How tall was Terry's gladiolus on Monday?

____

Mixed Applications

Solve.

CHOOSE A STRATEGY

- **Work Backward** • **Write a Number Sentence** • **Guess and Check** • **Make an Organized List**

5. Morgan's class reads every day in class. On Mondays, Wednesdays, and Fridays they read for 1 hour each day. On Tuesdays and Thursdays they read for $\frac{1}{2}$ hour each day. How many hours do they read each week?

____

6. Lisa gave 10 sheets of graph paper to Ted and 8 sheets to Nancy. Then Jill gave Lisa 12 sheets of graph paper. Lisa now has 35 sheets. How many sheets did she have to begin with?

____

Name ______________________________

LESSON 19.1

Estimating Sums and Differences

Use the number lines to estimate whether the fraction is closest to 0, to $\frac{1}{2}$, or to 1.

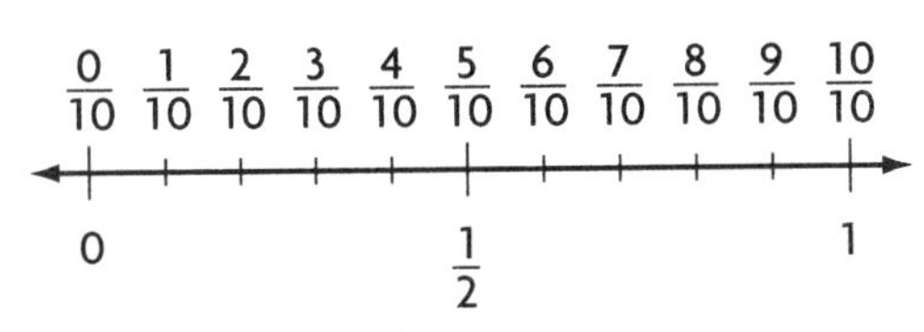

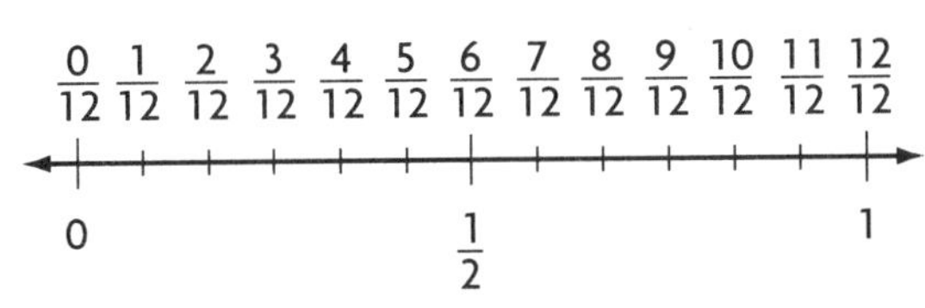

1. $\frac{4}{10}$ ______

2. $\frac{11}{12}$ ______

3. $\frac{2}{10}$ ______

4. $\frac{7}{12}$ ______

Write whether the fraction is closest to 0, to $\frac{1}{2}$, or to 1. You may use a number line.

5. $\frac{7}{8}$ ______

6. $\frac{3}{8}$ ______

7. $\frac{2}{9}$ ______

8. $\frac{1}{8}$ ______

Estimate each sum or difference.

9. $\frac{1}{2} + \frac{3}{4}$ ______

10. $\frac{1}{2} + \frac{5}{8}$ ______

11. $\frac{1}{4} + \frac{5}{9}$ ______

12. $\frac{6}{8} + \frac{2}{4}$ ______

13. $\frac{11}{12} - \frac{1}{9}$ ______

14. $\frac{5}{6} - \frac{3}{5}$ ______

15. $\frac{8}{9} - \frac{3}{4}$ ______

16. $\frac{7}{9} - \frac{5}{8}$ ______

Mixed Applications

17. Joanna talked on the phone for $\frac{3}{5}$ hour before dinner. She talked on the phone for $\frac{7}{8}$ hour after dinner. About how long did she talk in all?

____

18. Yolanda runs 3 miles a day, and Rob runs 2 miles a day. They run every day in March. How many miles do they run in all during March?

____

Name ___________________________

LESSON 19.2

Adding and Subtracting Like Fractions

Find the sum. Write the answer in simplest form.

1. $\frac{5}{7} + \frac{1}{7} = n$ ________

2. $\frac{4}{9} + \frac{3}{9} = n$ ________

3. $\frac{4}{12} + \frac{8}{12} = n$ ________

4. $\frac{3}{11} + \frac{7}{11} = n$ ________

5. $\frac{2}{8} + \frac{4}{8} = n$ ________

6. $\frac{7}{15} + \frac{4}{15} = n$ ________

7. $\frac{5}{9} + \frac{1}{9} = n$ ________

8. $\frac{1}{4} + \frac{2}{4} = n$ ________

Find the difference. Write the answer in simplest form.

9. $\frac{4}{7} - \frac{2}{7} = n$ ________

10. $\frac{3}{5} - \frac{1}{5} = n$ ________

11. $\frac{6}{12} - \frac{2}{12} = n$ ________

12. $\frac{3}{4} - \frac{2}{4} = n$ ________

13. $\frac{7}{9} - \frac{2}{9} = n$ ________

14. $\frac{4}{6} - \frac{1}{6} = n$ ________

15. $\frac{3}{8} - \frac{2}{8} = n$ ________

16. $\frac{9}{10} - \frac{5}{10} = n$ ________

Mixed Applications

17. George ran $\frac{3}{8}$ mile on Sunday and $\frac{2}{8}$ mile on Monday. How much farther did George run on Sunday than on Monday?

____

18. Lona pulled the wagon for $\frac{4}{10}$ hour. Eric pulled the wagon for $\frac{1}{10}$ hour. For how long did they pull the wagon in all?

____

19. Mary and her friends went to the amusement park. They left home at 8:30 A.M. They drove 1 hour 15 minutes to the park. They spent 6 hours at the park. Then they drove 1 hour 15 minutes home. What time was it when they returned home?

____

20. Josh had a great time at the amusement park with Mary and all their friends. Josh spent \$10.75 on souvenirs, \$4.50 on lunch, and \$3.00 on gas. How much money did he have left from his \$20 bill?

____

Name ____________________

LESSON 19.3

Adding and Subtracting Unlike Fractions

Find the sum or difference. Write the answer in simplest form.

1. $\frac{1}{2} + \frac{2}{8} = n$ ________
2. $\frac{2}{5} + \frac{1}{3} = n$ ________
3. $\frac{6}{8} + \frac{1}{4} = n$ ________
4. $\frac{9}{12} - \frac{2}{4} = n$ ________
5. $\frac{8}{16} - \frac{2}{8} = n$ ________
6. $\frac{2}{10} + \frac{3}{5} = n$ ________
7. $\frac{7}{9} - \frac{1}{3} = n$ ________
8. $\frac{4}{15} + \frac{2}{3} = n$ ________
9. $\frac{3}{8} - \frac{1}{4} = n$ ________
10. $\frac{6}{12} - \frac{2}{6} = n$ ________
11. $\frac{9}{10} - \frac{4}{5} = n$ ________
12. $\frac{6}{8} - \frac{1}{2} = n$ ________
13. $\frac{5}{8} + \frac{5}{16} = n$ ________
14. $\frac{4}{5} + \frac{1}{10} = n$ ________
15. $\frac{5}{9} - \frac{7}{18} = n$ ________
16. $\frac{1}{2} - \frac{3}{14} = n$ ________
17. $\frac{2}{20} + \frac{4}{5} = n$ ________
18. $\frac{1}{3} - \frac{2}{9} = n$ ________
19. $\frac{2}{6} - \frac{5}{18} = n$ ________
20. $\frac{3}{8} + \frac{2}{4} = n$ ________

Mixed Applications

21. Jade swam $\frac{1}{2}$ mile on Monday. On Wednesday she swam $\frac{3}{8}$ mile. How many miles did Jade swim in all?

____

22. Mike has 6 photo albums. Each photo album has 36 pictures in it. How many pictures does Mike have in all?

____

23. Monty spent $\frac{4}{5}$ hour mowing his lawn. Then he spent $\frac{2}{10}$ hour mowing his neighbor's lawn. How much longer did it take Monty to mow his lawn than his neighbor's lawn?

____

24. Laura bought 24 bananas. She gave 12 of them to her sister and 8 of them to her brother. What fraction of the bananas did Laura have left?

____

Name ____________________

LESSON 19.4

Practicing Addition and Subtraction

Find the sum or difference. Write each answer in simplest form.

1. $\frac{3}{6} - \frac{1}{6} = n$ ______

2. $\frac{5}{8} + \frac{3}{16} = n$ ______

3. $\frac{1}{4} - \frac{2}{12} = n$ ______

4. $\frac{8}{14} - \frac{7}{14} = n$ ______

5. $\frac{8}{9} - \frac{4}{9} = n$ ______

6. $\frac{4}{15} + \frac{3}{5} = n$ ______

7. $\frac{5}{8} - \frac{3}{16} = n$ ______

8. $\frac{3}{7} - \frac{2}{14} = n$ ______

9. $\frac{6}{20} + \frac{2}{10} = n$ ______

10. $\frac{1}{9} + \frac{2}{3} = n$ ______

11. $\frac{1}{7} - \frac{2}{35} = n$ ______

12. $\frac{5}{7} + \frac{1}{7} = n$ ______

13. $\frac{3}{8} + \frac{5}{16} = n$ ______

14. $\frac{7}{12} - \frac{1}{12} = n$ ______

15. $\frac{3}{5} + \frac{1}{4} = n$ ______

16. $\frac{7}{10} - \frac{2}{5} = n$ ______

Mixed Applications

17. Deena made $\frac{2}{5}$ pound of spaghetti. She made $\frac{3}{4}$ pound of meatballs. How much more meatballs did Deena make than spaghetti?

______

18. Roy had a box of popcorn. He ate $\frac{2}{3}$ of the popcorn, and his friends ate $\frac{2}{9}$ of the popcorn. What fraction of the box of popcorn is gone?

______

19. Harry, Jane, and Richard went to the mall. They started with $64.00. When they finished, they had $23.75 left. How much money did they spend at the mall?

______

20. Ed made 36 cookies on Monday. On Tuesday he made 29 cookies. He put 3 nuts on each cookie. How many nuts did Ed use on the cookies in all?

______

Name ______________________

LESSON 19.5

Choosing Addition or Subtraction

Tell whether you would add or subtract to solve the problem. Solve.

1. Jenna practiced piano $\frac{1}{6}$ hour on Friday, $\frac{1}{2}$ hour on Saturday, and $\frac{2}{3}$ hour on Sunday. How long did Jenna practice in all?

____

2. Ellen ate $\frac{1}{8}$ of the pizza. Leslie ate $\frac{1}{4}$ of the pizza, and Allen ate $\frac{1}{2}$ of the pizza. What was the total amount of pizza eaten?

____

3. Matthew lives $\frac{3}{4}$ mile from school. Evan lives $\frac{3}{12}$ mile from school. How much farther does Matthew live from school than Evan?

____

4. Morgan spent $\frac{2}{3}$ of his money on models and $\frac{2}{9}$ of his money on school supplies. How much more of his money did he spend on models than on school supplies?

____

Mixed Applications

5. Alison gave $\frac{1}{6}$ of her hair ribbons to her sister. She gave $\frac{2}{3}$ of the ribbons to her friend. What fraction of the ribbons did Alison give away in all?

____

6. Colin spent $\frac{3}{8}$ of his paycheck on a backpack and $\frac{1}{2}$ of his paycheck on shoes. Did he spend more money for the backpack or for the shoes?

____

7. Al has 12 model airplanes. Steve has 3 times as many model airplanes as Al has. How many airplanes does Steve have? How many airplanes do they have in all?

____

8. Keesha lives $\frac{3}{10}$ mile from the soccer field. Sandy lives $\frac{4}{5}$ mile from the field. How much closer to the field does Keesha live than Sandy?

____

9. Mr. Reed is taking the 24 students in his class on a field trip. Each student has to pay $7 to go. How much money is Mr. Reed's class paying to go on the field trip?

____

10. Jim saw 10 blue jays, 12 robins, and 4 crows in his backyard. What fraction of the birds he saw were robins?

____

Name ___________________________

Problem-Solving Strategy

Draw a Diagram

Draw a diagram to solve.

1. Some Upper Crest Middle School students are taking a field trip. Of the students going, $\frac{1}{2}$ are sixth graders, $\frac{1}{4}$ are seventh graders, and $\frac{1}{4}$ are eighth graders. There are 20 seventh graders going. How many students are going on the field trip?

   ____

2. Brett rides his bike a set number of miles every week. On Sunday he rides $\frac{2}{5}$ of his total distance. On Tuesday he rides $\frac{1}{2}$ of the distance, and on Thursday he rides $\frac{1}{10}$ of the distance. He rides 7 miles on Thursday. How many miles does he ride for the week?

   ____

3. Doug collects mugs. In his mug collection, $\frac{1}{8}$ are clear, $\frac{1}{2}$ are black, and $\frac{3}{8}$ are multicolored. Doug has 6 clear mugs. How many mugs does Doug have in all?

   ____

4. Ralph, Melissa, and Peter are comparing their ages. At 36 years old, Ralph is 8 years older than Peter. Peter is 6 years younger than Melissa. How old is Melissa?

   ____

Mixed Applications

Solve.

CHOOSE A STRATEGY

- **Draw a Diagram**
- **Use a Table**
- **Write a Number Sentence**
- **Guess and Check**

5. Steve and Ellen bought 2 popcorns, 2 sodas, and 1 hot dog. How much did they spend in all?

Food	Price
Popcorn	$2.25
Soda	$1.00
Hot dog	$1.50

   ____

6. David has 50 mugs to take to the craft fair. If he carries them in boxes that hold 12 mugs, how many boxes will he take?

   ____

Name ______________________

LESSON 20.1

Estimating Sums and Differences

Round the mixed number to the nearest $\frac{1}{2}$ or whole number. You may use a number line or a ruler.

1. $2\frac{1}{4}$ ______
2. $3\frac{1}{9}$ ______
3. $4\frac{2}{3}$ ______
4. $5\frac{7}{16}$ ______
5. $7\frac{11}{12}$ ______
6. $3\frac{5}{8}$ ______
7. $6\frac{3}{4}$ ______
8. $4\frac{7}{16}$ ______
9. $8\frac{1}{3}$ ______
10. $2\frac{7}{8}$ ______
11. $7\frac{2}{8}$ ______
12. $9\frac{4}{9}$ ______

Estimate the sum or difference.

13. $3\frac{1}{4} + 2\frac{7}{8}$ ______
14. $2\frac{1}{2} - 1\frac{3}{5}$ ______
15. $5\frac{7}{12} + 3\frac{1}{8}$ ______
16. $7\frac{1}{9} - 1\frac{15}{16}$ ______
17. $8\frac{2}{3} + 4\frac{3}{16}$ ______
18. $6\frac{2}{9} + 4\frac{1}{12}$ ______
19. $8\frac{9}{10} - 5\frac{1}{5}$ ______
20. $7\frac{14}{16} - 2\frac{8}{9}$ ______
21. $9\frac{2}{8} + 3\frac{5}{12}$ ______

Mixed Applications

22. Mary earns $5.25 an hour baby-sitting for her neighbors. Last month she worked 87 hours. How much money did she make?

______

23. Franklin can type about 35 words per minute. How long will it take him to type 1,575 words?

______

24. Sara buys $2\frac{5}{8}$ feet of cloth and $3\frac{1}{4}$ feet of cloth. About how many feet of cloth does she buy in all?

______

25. Jack is making a project out of wood. He started with a board $3\frac{1}{2}$ feet long. He has used $2\frac{15}{16}$ feet already. About how much does he have left?

______

Name ____________________

LESSON 20.2

Adding Mixed Numbers

Find the sum. You may wish to draw a picture.

1. $2\frac{3}{8} + 3\frac{1}{4}$

2. $4\frac{1}{3} + 3\frac{1}{6}$

3. $1\frac{5}{12} + 2\frac{1}{6}$

4. $3\frac{5}{8} + 3\frac{3}{4}$

5. $1\frac{1}{10} + 4\frac{2}{5}$

6. $3\frac{1}{9} + 4\frac{1}{3}$

7. $2\frac{3}{5} + 5\frac{7}{10}$

8. $4\frac{1}{12} + 2\frac{1}{3}$

9. $7\frac{2}{3} + 1\frac{5}{12} = n$ ________

10. $2\frac{4}{5} + 5\frac{3}{10} = n$ ________

11. $3\frac{1}{4} + 3\frac{7}{8} = n$ ________

12. $2\frac{2}{3} + 4\frac{1}{6} = n$ ________

13. $1\frac{1}{2} + 3\frac{5}{6} = n$ ________

14. $4\frac{2}{3} + 3\frac{5}{6} = n$ ________

15. $5\frac{5}{12} + 2\frac{1}{6} = n$ ________

16. $8\frac{2}{9} + 1\frac{1}{3} = n$ ________

Mixed Applications

17. Tim and Ralph are making a tower. They each built a separate section. Tim's section was $4\frac{7}{8}$ feet tall, and Ralph's section was $3\frac{1}{2}$ feet tall. How tall will the tower be when they join the sections?

____

18. At the last fifth-grade assembly, there were 5 classes with 31 students and 4 classes with 28 students. How many students attended the assembly?

____

19. Harriet and Felicia worked for the local charity on the weekend. Harriet worked $5\frac{1}{6}$ hours, and Felicia worked $3\frac{1}{4}$ hours. How many hours did the girls work together for the charity?

____

20. Jane's family traveled 171 miles on Monday and 377 miles on Tuesday. Philip's family traveled 396 miles on Monday and 211 miles on Tuesday. Whose family traveled more miles?

____

Name ______________________________

Subtracting Mixed Numbers

Subtract. Write the answer in simplest form.

1. $3\frac{7}{10} = 3\frac{7}{10}$
   $-1\frac{2}{5} = 1\frac{4}{10}$

2. $5\frac{3}{4} = 5\frac{6}{8}$
   $-2\frac{1}{8} = 2\frac{1}{8}$

3. $8\frac{5}{6} = 8\frac{10}{12}$
   $-2\frac{1}{12} = 2\frac{1}{12}$

4. $7\frac{1}{2} = 7\frac{3}{6}$
   $-4\frac{1}{6} = 4\frac{1}{6}$

5. $9\frac{9}{10} = 9\frac{9}{10}$
   $-4\frac{3}{5} = 4\frac{6}{10}$

6. $5\frac{4}{9} = 5\frac{4}{9}$
   $-3\frac{1}{3} = 3\frac{3}{9}$

7. $7\frac{3}{8} - 2\frac{1}{4} = n$ ________

8. $5\frac{4}{5} - 3\frac{3}{10} = n$ ________

9. $3\frac{5}{12} - 2\frac{1}{4} = n$ ________

10. $5\frac{7}{12} - 3\frac{1}{2} = n$ ________

11. $9\frac{5}{6} - 4\frac{2}{3} = n$ ________

12. $4\frac{7}{8} - 2\frac{3}{4} = n$ ________

13. $6\frac{3}{4} - 4\frac{2}{8} = n$ ________

14. $3\frac{3}{4} - 2\frac{5}{8} = n$ ________

15. $4\frac{11}{12} - \frac{7}{12} = n$ ________

16. $5\frac{1}{2} - 3\frac{3}{8} = n$ ________

Mixed Applications

17. Sam made the chart at the right to keep track of how much wood he had for projects. He forgot to enter some of the numbers. Complete the table.

Type of Wood	Started With	Feet Used	Feet Left
Oak	$15\frac{1}{2}$	$9\frac{1}{4}$	____
Pine	$22\frac{5}{8}$	____	$10\frac{1}{4}$
Maple	____	$12\frac{3}{4}$	$2\frac{1}{6}$
Cherry	$20\frac{3}{4}$	$5\frac{3}{8}$	____

18. Each week Henry will work $3\frac{1}{2}$ hours on Wednesday and $4\frac{1}{4}$ hours on Friday. How many hours will he work each week?

____

Name ______________________________

LESSON 20.3

Problem-Solving Strategy

Work Backward

Work backward to solve.

1. Emily used wallpaper border to outline her window. She used $6\frac{1}{3}$ yards to outline the door and $1\frac{1}{6}$ yards to outline a shelf. She used $9\frac{1}{2}$ yards of border in all. How much border did she use on the window?

____

2. On Friday Jake had done a total of 125 push-ups in five days. He did 20 on Monday, 30 on Tuesday, 15 on Wednesday, and 20 on Thursday. How many push-ups did he do on Friday?

____

3. Dirk spent $3\frac{3}{4}$ hours outside on Saturday. During that time he spent $1\frac{1}{2}$ hours at the park and $1\frac{1}{4}$ hours in a friend's yard. He also rode his bicycle. How much time did he spend riding his bicycle?

____

4. Terry saved $60 to spend on a party for her mother. She spent $25 for a cake and $12 for party decorations. She spent the rest on a gift. How much did she spend on the gift?

____

Mixed Applications

Solve.

CHOOSE A STRATEGY

- **Work Backward**
- **Draw a Diagram**
- **Act it Out**
- **Write a Number Sentence**

5. Marlinda bought 32 inches of butcher paper for her project. She used $15\frac{1}{4}$ inches for the background and $12\frac{3}{4}$ inches for different parts of the project. How much butcher paper did she have left?

____

6. Ingrid planted a garden. In the garden $\frac{1}{2}$ of the rows are tomatoes, $\frac{1}{4}$ of the rows are green beans, and $\frac{2}{8}$ of the rows are lettuce. She has 5 rows of green beans. How many rows are in the garden?

____

Name ______________________________

Subtracting Mixed Numbers

Match the mixed number with the fraction bars.

1. $2\frac{1}{4}$ ______ 2. $2\frac{3}{4}$ ______ 3. $3\frac{3}{8}$ ______

a. 1 | $\frac{1}{4}$ $\frac{1}{4}$ $\frac{1}{4}$ $\frac{1}{4}$ $\frac{1}{4}$ $\frac{1}{4}$ $\frac{1}{4}$

b. 1 | 1 | $\frac{1}{8}$ $\frac{1}{8}$ $\frac{1}{8}$ $\frac{1}{8}$ $\frac{1}{8}$ $\frac{1}{8}$ $\frac{1}{8}$ $\frac{1}{8}$ $\frac{1}{8}$ $\frac{1}{8}$ $\frac{1}{8}$

c. 1 | $\frac{1}{4}$ $\frac{1}{4}$ $\frac{1}{4}$ $\frac{1}{4}$ $\frac{1}{4}$

Use fraction bars to find the difference.

4. $3\frac{2}{3} - \frac{1}{6}$

5. $7\frac{1}{4} - 3\frac{3}{8}$

6. $4\frac{3}{10} - 2\frac{4}{5}$

7. $6\frac{2}{3} - 4\frac{5}{6}$

8. $8\frac{1}{2} - 1\frac{5}{6}$

9. $3\frac{1}{8} - 1\frac{1}{2}$

10. $7\frac{1}{10} - 4\frac{2}{5}$

11. $10\frac{3}{8} - 5\frac{3}{4}$

12. $6\frac{11}{12} - 2\frac{2}{3} = n$ __________

13. $4\frac{1}{5} - 1\frac{7}{10} = n$ __________

14. $5\frac{5}{8} - 1\frac{3}{4} = n$ __________

15. $5\frac{1}{2} - 2\frac{7}{12} = n$ __________

16. $8\frac{1}{6} - 4\frac{5}{12} = n$ __________

17. $7\frac{1}{4} - 6\frac{7}{12} = n$ __________

Mixed Applications

18. Stacey had $3\frac{3}{4}$ cakes for her party. She had $\frac{1}{8}$ of a cake left after the party. How much cake did Stacey use at her party?

____

19. Martha spent $2\frac{1}{2}$ hours reading on Saturday. She spent $\frac{3}{4}$ of an hour reading on Sunday. How many hours did she spend reading this weekend?

____

Name ______________________________

LESSON 21.1

Precise Measurements

Vocabulary

Fill in the blank.

1. ________________ means finding a unit that measures nearest to the actual length of an object.

For Problems 2–3, use a customary ruler.

2. Measure the length of the paper clip to the nearest $\frac{1}{8}$ inch.

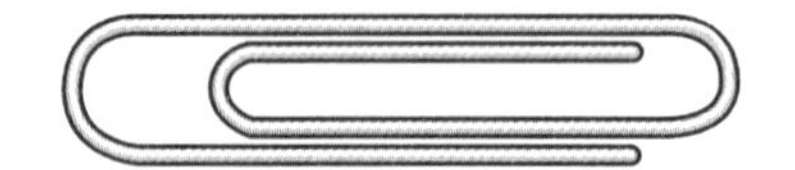

____

3. Measure the length of the crayon to the nearest $\frac{1}{16}$ inch.

____

Draw a line to the given length.

4. $1\frac{3}{4}$ inch

5. $2\frac{3}{16}$ inch

6. $3\frac{5}{16}$ inch

Use a ruler to compare the measures. Write $<$, $>$, or $=$ in each ○.

7. $4\frac{3}{16}$ in. ○ $3\frac{15}{16}$ in.

8. $3\frac{6}{8}$ in. ○ $3\frac{3}{4}$ in.

9. $2\frac{3}{8}$ in. ○ $2\frac{3}{16}$ in.

Mixed Applications

10. Karina's art teacher gave her an $8\frac{1}{2}$-inch by 11-inch piece of paper. He told her to leave a $\frac{3}{4}$-inch margin on all 4 sides. What are the dimensions of the remaining area?

____

11. Eliza measures her hair ribbon. It is $9\frac{10}{16}$ inch long. Mindy's hair ribbon is $9\frac{5}{8}$ inch long. Who has the longer hair ribbon?

____

____

Name ____________________

LESSON 21.2

Changing Customary Units

Write *multiply* or *divide* to tell how to change the unit.

1. feet to inches __________

2. miles to yards __________

3. feet to yards __________

4. feet to miles __________

Change the unit. You may use a calculator.

5. 3 ft = ______ in.

6. 4 yd = ______ ft

7. 60 in. = ______ ft

8. 3 yd = ______ in.

9. 36 ft = ______ yd

10. 1,760 yd = ______ mi

Write *more* or *fewer*.

11. When I change yards to feet, I expect to have ________ feet.

12. When I change yards to miles, I expect to have ________ miles.

Write *multiply* or *divide*. Solve.

13. How many yards are in 2 miles?

____

14. How many yards are in 126 feet?

____

Mixed Applications

For Problems 15–17, use the diagram at the right.

Mr. Rogers's Classroom

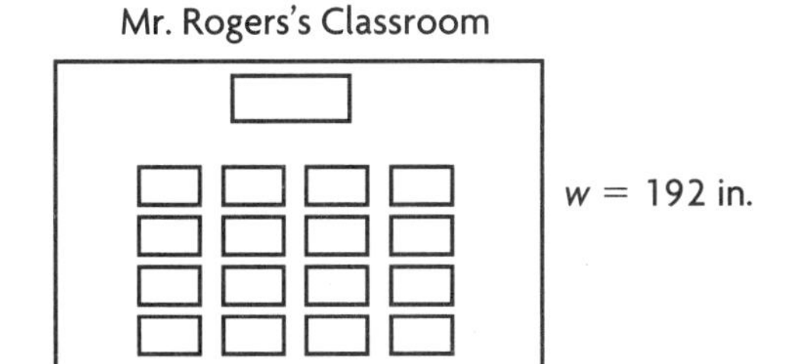

15. Mr. Rogers is decorating his classroom for a year-end party. He is measuring the width of his room. How many feet is the width?

____

16. Mr. Rogers plans to hang crepe paper across the front of the classroom. Should he buy the 5-foot, 15-foot, or 30-foot length?

____

17. In feet, what are the area and perimeter of Mr. Rogers's classroom?

____

____

Name ______________________________

Computing Customary Units

Rename the measurements.

1. 45 in. = ______ ft ______ in.
2. 14 ft = ______ yd ______ ft
3. 8 ft 19 in. = 9 ft ______ in.
4. 3 ft 3 in. = 2 ft ______ in.
5. 9 yd 2 ft = 8 yd ______ ft
6. 13 yd 4 ft = 14 yd ______ ft

Find the sum or difference.

7. 6 ft 5 in. + 3 ft 9 in.

8. 8 ft 10 in. + 2 ft 7 in.

9. 3 yd 2 ft + 4 yd 2 ft

10. 7 ft 3 in. − 3 ft 6 in.

11. 9 yd 7 ft − 6 yd 8 ft

12. 12 ft 4 in. − 6 ft 11 in.

Mixed Applications

13. Cobb buys $2\frac{1}{2}$ yards of striped fabric and $3\frac{1}{3}$ yards of printed fabric. How much fabric does Cobb buy in all?

____

14. Dolores is counting the change in her drawer. When she gets 6 more nickels, she will have $5 in nickels. How many nickels does she have now?

____

15. Nasira is buying ribbon to frame the perimeter of a wall hanging that is 50 inches long and 60 inches wide. If ribbon is sold by the yard, how many yards does she need?

____

16. Masha and Lu are cutting pieces of rope. Masha cuts 4 feet 5 inches of rope. Lu cuts 3 feet 17 inches of rope. Who has the larger piece of rope?

____

____

Name ____________________

Capacity

Change the unit.

1. 16 pt = ______ gal

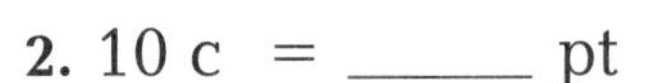

2. 10 c = ______ pt

3. 4 qt = ______ c

4. 1 gal = ______ c

5. 32 fl oz = ______ pt

6. 4 qt = ______ pt

7. 16 qt = ______ gal

8. 8 c = ______ fl oz

For Exercises 9–12, use the picture.

9. How many pints of ice cream?

____

10. How many cups of soda?

____

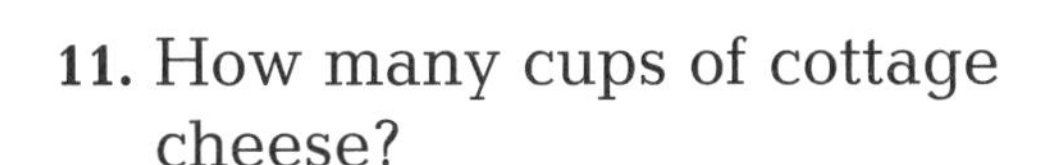

11. How many cups of cottage cheese?

____

12. How many pints of syrup?

____

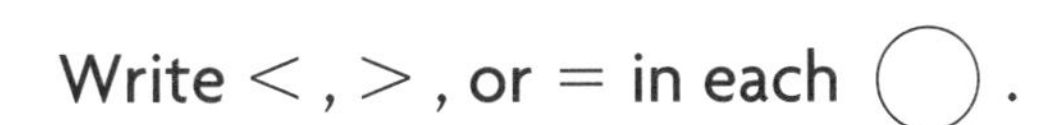

Write <, >, or = in each ○.

13. 3 gal ○ 14 qt

14. 1 qt ○ 32 fl oz

15. 5 c ○ 32 fl oz

16. 132 fl oz ○ 1 gal

17. 5 qt ○ 20 c

18. 8 pt ○ 3 qt

Mixed Applications

19. The water cooler in Juan-Carlos's office uses 5-gallon bottles of water. How many cups can each bottle fill?

____

20. Sharon is bringing orange juice to the class party. She needs enough juice to fill 24 cups. How many quarts of juice should she buy?

____

Name ______________________

LESSON 21.4

Weight

What unit would you use to describe the weight of these objects? Write *tons*, *pounds*, or *ounces*.

1. a 20-inch TV

______

2. a hamburger

______

3. a 747 airplane

______

Write *more* or *fewer* for each statement.

4. When I change tons to ounces, I expect to have

______ ounces.

5. When I change ounces to pounds, I expect to have

______ pounds.

Write *multiply* or *divide*. Change the unit. You may use a calculator.

6. 30 lb = __?__ oz

______

7. 32,000 oz = __?__ T

______

8. 3 T = __?__ lb

______

Write which one is heavier.

9. 40 oz or 2 lb

______

10. 3,000 lb or 2 T

______

11. $\frac{1}{2}$ T or 1,500 lb

______

Mixed Applications

12. At Wally's, a 12-ounce box of sunflower seeds sells for $2.59. At Barney's, sunflower seeds sell for $3.14 per pound. Which is the better buy?

______

13. Fred has 36 feet of fencing to build a pen for his pet duck. What dimensions should he make the pen to get the greatest possible area?

______

Name ______________________

LESSON 21.5

Elapsed Time

Write the elapsed time for each hike.

1.

Time the hike started

Time the hike finished

Elapsed time ____________

2.

Time the hike started

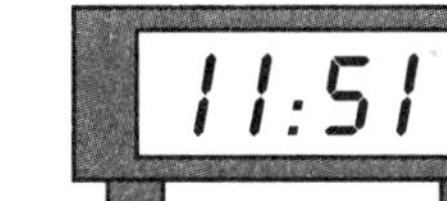

Time the hike finished

Elapsed time ____________

Look at Nadeem's schedule for the day and complete the table.

	Activity	Starting Time	Ending Time	Elapsed Time
3.	School	8:05 A.M.	2:25 P.M.	________
4.	Baseball practice	2:45 P.M.	________	2 hr 35 min
5.	Dinner	________	6:34 P.M.	31 min
6.	Homework	6:38 P.M.	7:57 P.M.	________

Mixed Applications

For Problems 7–8, use the calendars.

7. Mr. Scopa's fifth-grade class is going on a 60-hour class trip. If they leave at 9 A.M. on June 3, when do they return?

____

8. An 8-week weekday summer camp begins on June 24. What is the final day of camp?

____

9. Julie bought 16 pounds of apples at $1.69 per pound. How much did Julie pay?

____

JUNE

Sun	Mon	Tue	Wed	Thu	Fri	Sat
						1
2	3	4	5	6	7	8
9	10	11	12	13	14	15
16	17	18	19	20	21	22
23/30	24	25	26	27	28	29

JULY

Sun	Mon	Tue	Wed	Thu	Fri	Sat
	1	2	3	4	5	6
7	8	9	10	11	12	13
14	15	16	17	18	19	20
21	22	23	24	25	26	27
28	29	30	31			

AUGUST

Sun	Mon	Tue	Wed	Thu	Fri	Sat
				1	2	3
4	5	6	7	8	9	10
11	12	13	14	15	16	17
18	19	20	21	22	23	24
25	26	27	28	29	30	31

Name ________________________________

Problem-Solving Strategy

Make a Table

Make a table to solve.

1. The pool at the community center is used from 6:00 A.M. until 8:30 A.M. by the swim team. Then there is a one-hour open swim followed by four 45-minute swim classes. At what time is the pool available again?

   ____

2. Tomás arrives at camp at 9:30 A.M. He has swimming for $1\frac{1}{2}$ hours, archery for 1 hour, and lunch for 30 minutes. Then he has crafts for $2\frac{1}{2}$ hours. At what time does Tomás finish crafts?

   ____

3. The youth symphony begins auditions at 10:00 A.M. Each student is given 10 minutes to perform. If Claudia is the 12th in line, at what time is her audition?

   ____

4. Kelly reads stories to children at the library. There are 3 sessions. Each lasts 45 minutes, with 30 minutes between sessions. If Kelly starts reading at 10:00 A.M., at what time does she finish?

   ____

Mixed Applications

Solve.

CHOOSE A STRATEGY

- **Work Backward** • **Make a Table** • **Guess and Check** • **Write a Number Sentence**

5. It is now 6:30 P.M. Jerry has just finished eating dinner, which took 40 minutes. Before that, he was at soccer practice for 2 hours and 45 minutes. At what time did soccer practice begin?

   ____

6. Holly exercises for 20 minutes a day for 3 weeks. Then she exercises for 30 minutes a day for 2 weeks. At the end of the fifth week, how many hours has Holly exercised?

   ____

7. Yoma bought a 35-ounce box of raisins for $3.79. Gloria paid $1.79 for a pound of raisins. Who got the better bargain?

   ____

8. Gil's mom changes the oil in her car every 3,000 miles. If she drives 18,000 miles per year, how many times does she change her oil each year?

   ____

Name ______________________________

Temperature Changes

Vocabulary

Fill in the blanks.

1. ______________ ______________ are customary units for measuring temperature.

2. ______________ ______________ are metric units for measuring temperature.

Find the difference in temperature.

3. the temperature of the oven, 78°C, and the temperature of the room, 20°C

   ____

4. the temperature at the indoor pool, 85°F, and the temperature outside, 37°F

   ____

5. the temperature at the mountain-top, ⁻6°F, and the temperature at the base, 37°F

   ____

6. the temperature in the freezer, ⁻12°C, and the temperature in the room, 23°C

   ____

Complete the table.

	Starting Temperature	Change in Temperature	Final Temperature
7.	6°C	rose 23°C	______
8.	73°F	______	27°F
9.	______	fell 26°C	⁻3°C
10.	58°F	______	75°F

Mixed Applications

11. At dusk the temperature was 37°F. The temperature had fallen 15°F by sunrise. At noon the temperature was 28°F warmer. What was the temperature at noon?

    ____

12. Lucas bikes $3\frac{1}{2}$ miles on the bike trail. Ben bikes $2\frac{3}{8}$ miles. How much farther does Lucas bike than Ben?

    ____

Name ____________________

LESSON 22.1

Multiplying Fractions and Whole Numbers

Write a number sentence for each picture.

1. 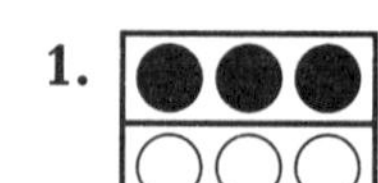
____

2. 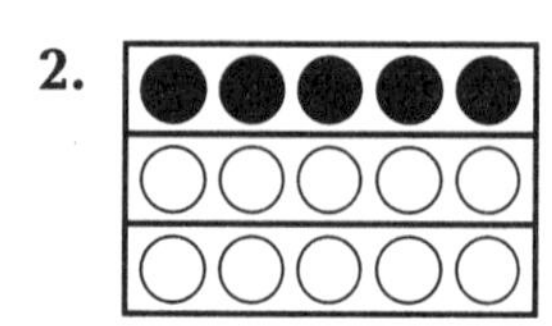
____

3. 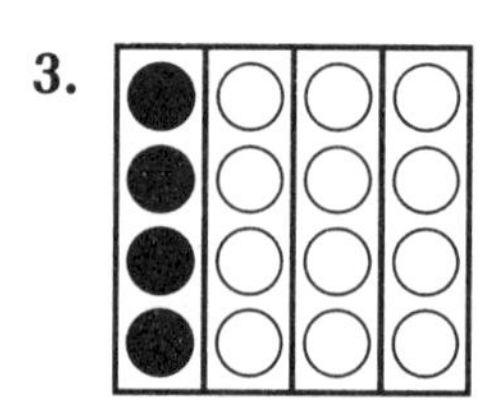
____

4. 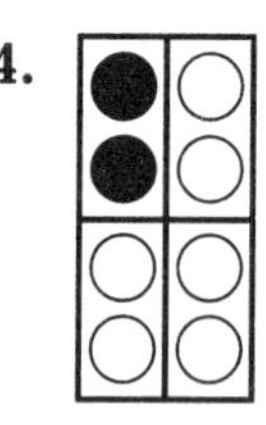
____

5. 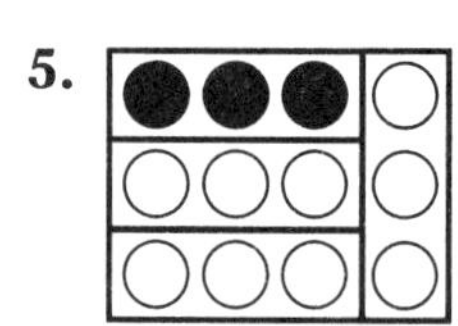
____

6. 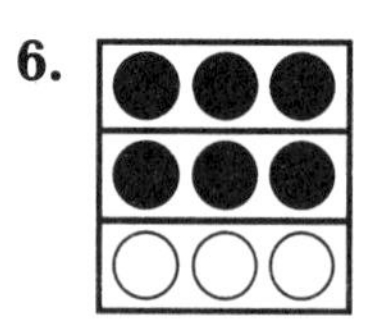
____

7. 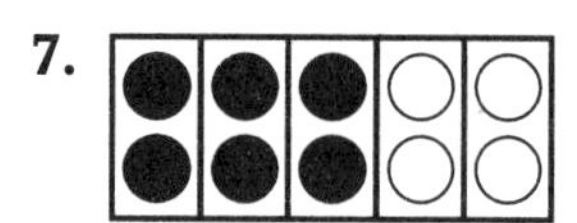
____

8. ____________

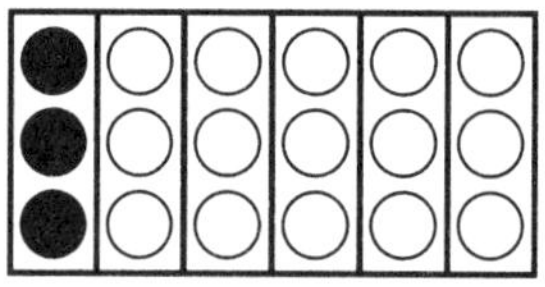

9. ____________

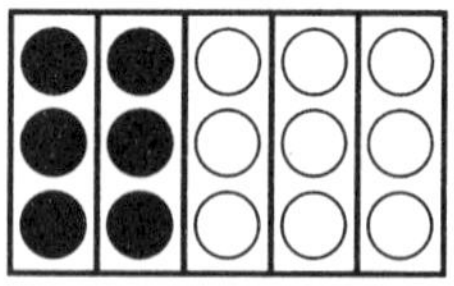

10. ____________

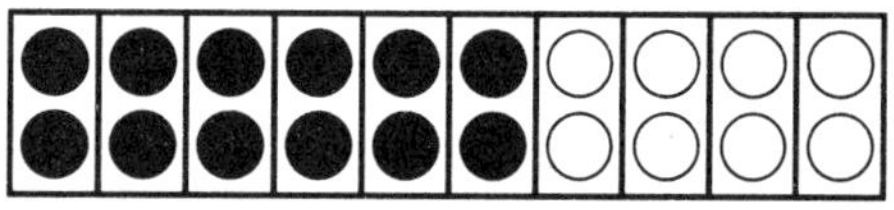

Find the product.

11. $\frac{1}{6} \times 18 =$ ______

12. $\frac{1}{7} \times 21 =$ ______

13. $\frac{1}{4} \times 16 =$ ______

14. $\frac{3}{8} \times 24 =$ ______

15. $\frac{2}{7} \times 14 =$ ______

16. $\frac{5}{8} \times 24 =$ ______

17. $12 \times \frac{3}{4} =$ ______

18. $24 \times \frac{5}{6} =$ ______

19. $18 \times \frac{7}{9} =$ ______

Mixed Applications

20. Kimberly has 20 books. Of those books, $\frac{2}{5}$ are mysteries. How many books are mysteries?

____

21. Dan walks to the store every week. It is $\frac{11}{12}$ mile from his house. If he has walked $\frac{3}{4}$ mile, how far must he still walk?

____

Name ______________________

Multiplying a Fraction by a Fraction

Make a paper-folding model to find the product.

1. $\frac{1}{3} \times \frac{1}{2} = n$ ______

2. $\frac{1}{4} \times \frac{1}{5} = n$ ______

3. $\frac{1}{4} \times \frac{3}{4} = n$ ______

Find the amount needed of each ingredient to cut the recipe in half.

Waffles

2 eggs	$\frac{1}{2}$ teaspoon baking soda	$\frac{3}{4}$ cup milk
$\frac{7}{8}$ cup flour	$\frac{1}{4}$ teaspoon salt	$\frac{2}{3}$ cup butter
2 teaspoons baking powder	2 tablespoons sugar	$\frac{1}{2}$ teaspoon vanilla
	$\frac{3}{4}$ cup sour cream	

4. eggs ______

5. flour ______

6. baking powder ______

7. baking soda ______

8. salt ______

9. sugar ______

10. sour cream ______

11. milk ______

12. butter ______

13. vanilla ______

Mixed Applications

14. Mary spent $\frac{8}{9}$ hour working on homework. She spent $\frac{1}{4}$ of that time on science. What part of an hour did she spend on science?

______

15. Bob combined $\frac{1}{3}$ gallon of orange juice, $\frac{3}{4}$ gallon of pineapple juice, and $\frac{5}{12}$ gallon of ginger ale to make punch. How many gallons did he make?

______

Name ______________________________

LESSON 22.3

More About Multiplying a Fraction by a Fraction

Write a number sentence for the picture.

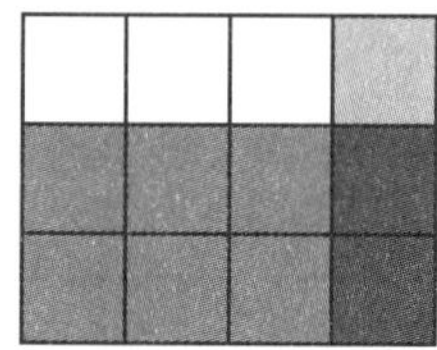

2. 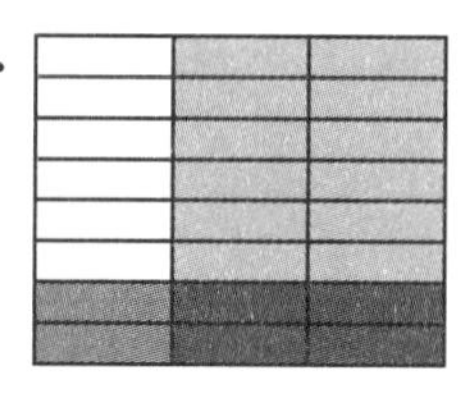

3. 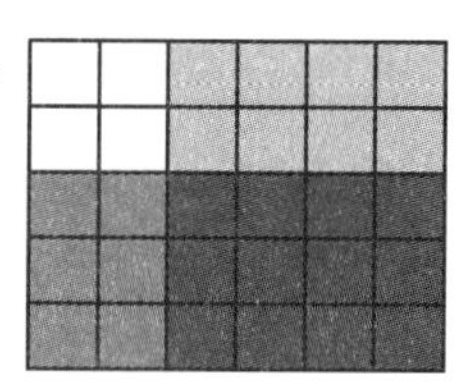

______________ ______________ ______________

Draw fraction squares to find the product.

4. $\frac{1}{3} \times \frac{1}{5} = n$ 5. $\frac{2}{5} \times \frac{1}{4} = n$ 6. $\frac{2}{3} \times \frac{1}{2} = n$ 7. $\frac{5}{6} \times \frac{2}{3} = n$

______________ ______________ ______________ ______________

Multiply. Write the answer in simplest form.

8. $\frac{1}{6} \times \frac{1}{3} = n$ 9. $\frac{2}{3} \times \frac{3}{5} = n$ 10. $\frac{1}{4} \times \frac{2}{7} = n$ 11. $\frac{4}{5} \times \frac{3}{8} = n$

______________ ______________ ______________ ______________

12. $\frac{1}{6} \times \frac{7}{8} = n$ 13. $\frac{3}{7} \times \frac{5}{8} = n$ 14. $\frac{11}{12} \times \frac{4}{9} = n$ 15. $\frac{7}{9} \times \frac{5}{6} = n$

______________ ______________ ______________ ______________

Mixed Applications

16. Sarah found $\frac{15}{16}$ cup of sugar in the cupboard. She used $\frac{2}{3}$ cup to make cookies. How much sugar was left?

____

17. Tim has 3 cups of cream. His recipe calls for 36 fluid ounces of cream. How much cream is left after Tim makes the recipe?

____

Name ______________________

LESSON 22.4

Multiplying Fractions and Mixed Numbers

Write a number sentence for the picture.

1. 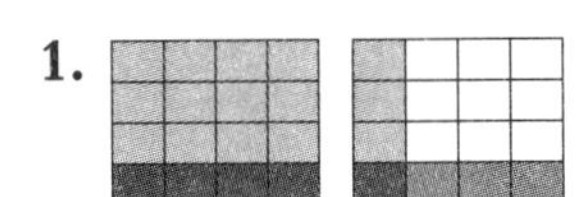
____

2. 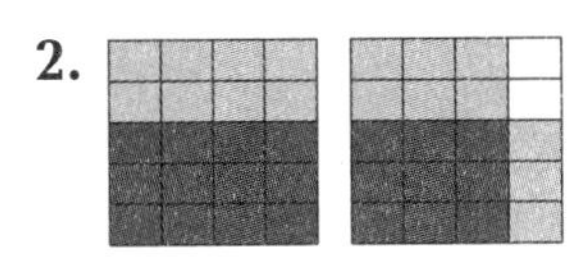
____

3. 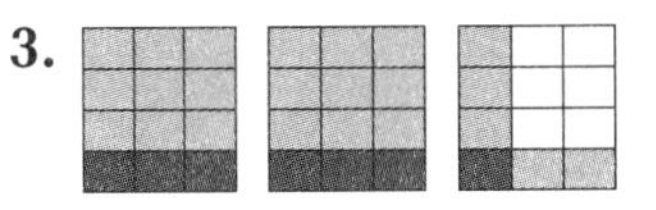
____

Draw fraction squares to help you find the product.

4. $\frac{2}{5} \times 1\frac{1}{3} = n$ ______________

5. $\frac{2}{3} \times 2\frac{1}{4} = n$ ______________

6. $\frac{3}{4} \times 3\frac{2}{3} = n$ ______________

Multiply. Write the answer in simplest form.

7. $\frac{1}{3} \times 2\frac{1}{4} = n$ ______________

8. $\frac{1}{6} \times 3\frac{1}{2} = n$ ______________

9. $\frac{2}{3} \times 1\frac{1}{2} = n$ ______________

10. $\frac{5}{6} \times 1\frac{2}{3} = n$ ______________

11. $\frac{3}{4} \times 2\frac{4}{5} = n$ ______________

12. $\frac{1}{3} \times 3\frac{2}{5} = n$ ______________

13. $\frac{2}{3} \times 2\frac{2}{3} = n$ ______________

14. $\frac{1}{2} \times 3\frac{5}{6} = n$ ______________

15. $\frac{3}{5} \times 1\frac{3}{4} = n$ ______________

Mixed Applications

16. On a recent trip, Mosi drove $2\frac{1}{2}$ hours on Monday, $1\frac{5}{6}$ hours on Tuesday, and $2\frac{2}{3}$ hours on Wednesday. How many hours did Mosi drive in all?

____

17. Martha had $3\frac{1}{3}$ dozen eggs. She used $\frac{3}{4}$ dozen to make cookies for a bake sale. How many dozen eggs does she have left?

____

Name ______________________________

Problem-Solving Strategy

Make a Model

Make a model to solve.

1. Casey has $3\frac{1}{5}$ yards of white yarn. Of that, $\frac{3}{4}$ is wool. How many yards of white wool yarn does Casey have?

   ____

2. Sam is taller than Brian. Betty is shorter than Sam, but taller than Susan. Susan is taller than Brian. Who is the shortest person?

   ____

3. A garden has an area of $12\frac{3}{4}$ square yards. Roses cover $\frac{1}{3}$ of the garden, and sunflowers cover $\frac{1}{4}$ of the garden. How many square yards do the roses cover? the sunflowers?

   ____

   ____

4. Toby found $2.29 in coins. Of those coins, 5 are quarters. He has four times as many nickels as quarters. The rest are pennies. How many of each coin does he have?

   ____

   ____

Mixed Applications

Solve.

CHOOSE A STRATEGY

- Find a Pattern
- Write a Number Sentence
- Guess and Check
- Work Backward
- Make a Model

5. Ann and Bob went to a movie. Together they had $27.60. The tickets cost $6.50 each. Each bought a drink for $2.30 and a small popcorn for $1.75. Movie trading cards cost $0.65 a pack. How many packs of trading cards could they buy?

   ____

6. A bus picks up 1 person at the first stop, 3 people at the second stop, 7 at the third stop, and 13 at the fourth stop. If this pattern continues, how many people will get on at the fifth stop? the sixth stop?

   ____

Name ______________________________

Line Relationships

Vocabulary

Write the correct letter from Column 2.

Column 1

_____ **1.** lines in a plane that never intersect and are the same distance from each other

_____ **2.** lines that intersect to form four right angles

_____ **3.** a flat surface with no end

_____ **4.** a part of a line

_____ **5.** lines that cross at one point

Column 2

a. perpendicular

b. plane

c. line segment

d. parallel

e. intersecting

For Exercises 6–9, use the figure.

6. Name the point where $\overline{AC}$ and $\overline{DC}$ intersect.

____

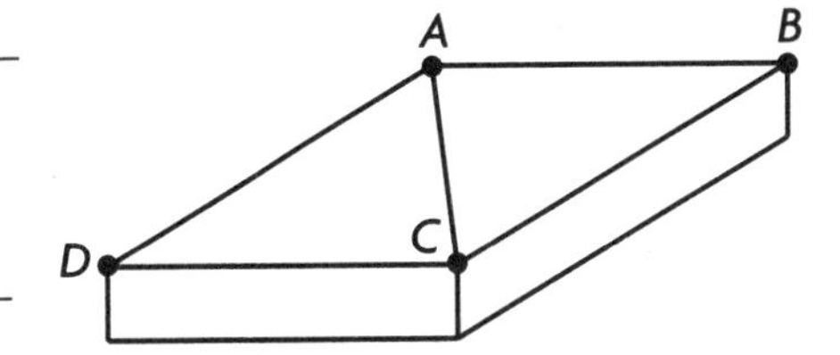

7. Name a line segment on plane *ABCD* that is perpendicular to $\overline{AB}$. ______________

8. Name a line segment that intersects but is not perpendicular to $\overline{AB}$. ______________

9. Name the line segment that is parallel to $\overline{AB}$.

____

Mixed Applications

10. In Carvertown, Main Street and James Street are perpendicular. What kind of angles are formed at their intersection?

____

11. Ben must drive 270 miles on Highway 5. He has now driven $\frac{1}{3}$ of the way. How many miles has he driven?

____

Name ____________________

LESSON 23.2

Rays and Angles

Vocabulary

Complete.

1. A ____________ is part of a line that has one endpoint and goes on forever in one direction.

2. An ____________ is formed when two rays have the same endpoint.

Identify the angle. Write *right*, *acute*, or *obtuse*.

3. 

4. 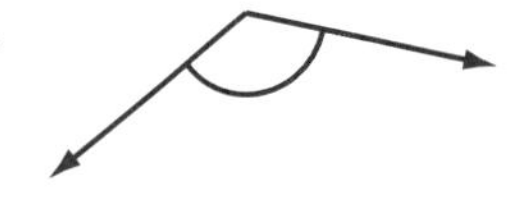

5. 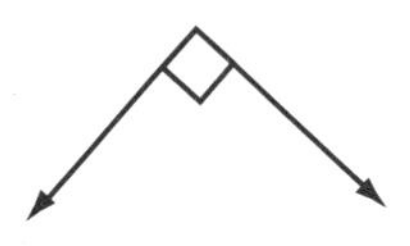

6. 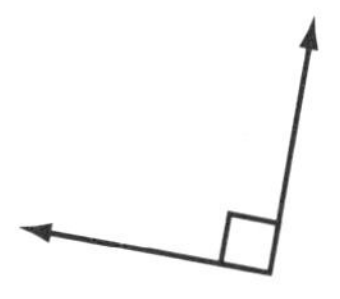

For Exercises 7–15, use the figure. Identify the angle. Write right, acute, or obtuse.

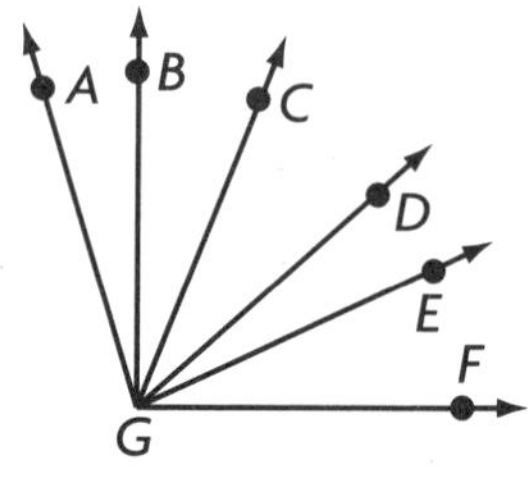

7. $\angle AGB$ ____________

8. $\angle CGB$ ____________

9. $\angle FGA$ ____________

10. $\angle AGF$ ____________

11. $\angle FGB$ ____________

12. $\angle EGF$ ____________

13. $\angle BGF$ ____________

14. $\angle EGB$ ____________

15. $\angle DGB$ ____________

Mixed Applications

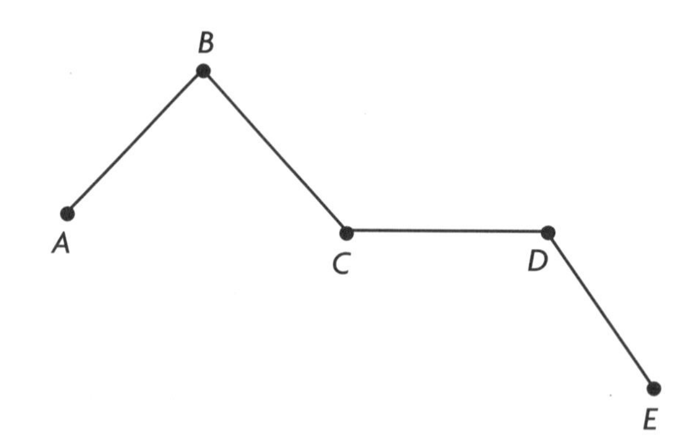

16. Luis is following a trail. If the points on the trail were labeled as shown, describe $\angle BCD$.

____

17. The trail is 35 miles long. Luis has hiked 7 miles. What fraction of the trail has he hiked?

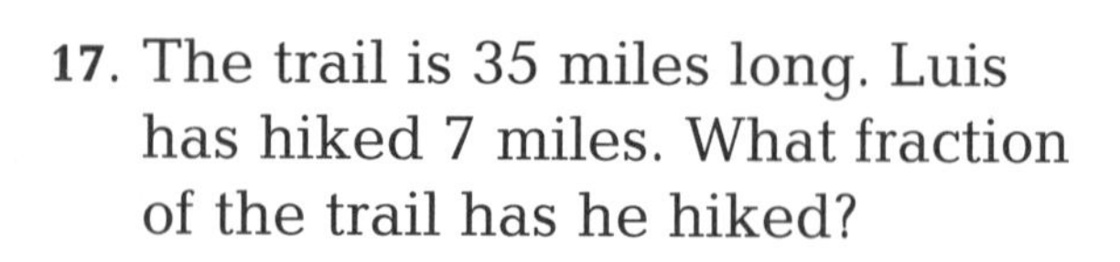

____

Name ______________________________

LESSON 23.3

Measuring Angles

Vocabulary

Complete.

1. The unit used to measure an angle is called a ____________.

2. A ____________ is a tool for measuring the size of the opening of an angle.

Use a protractor to measure each angle.

3. 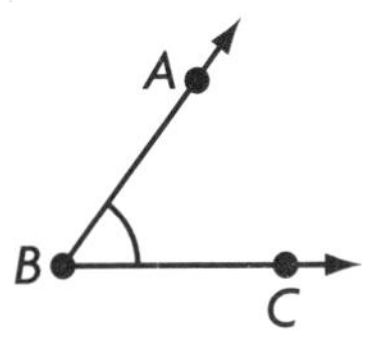

____

4. 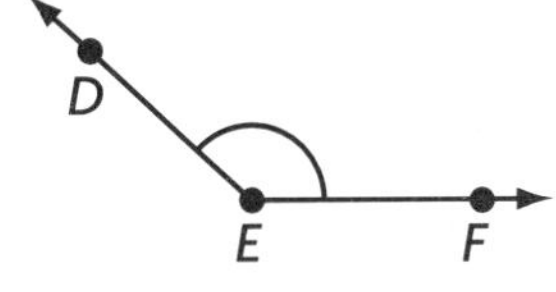

____

5. 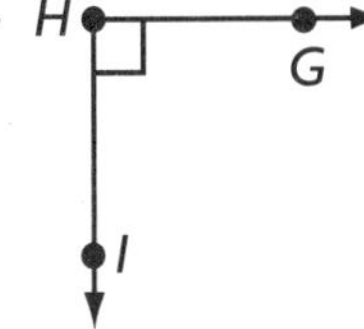

____

6. 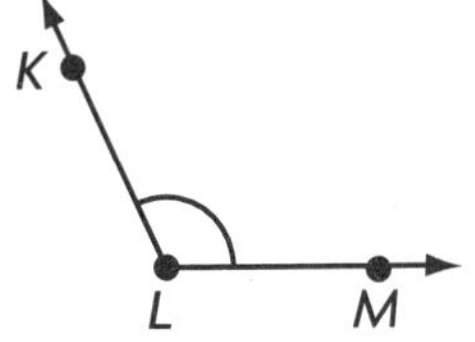

____

7. 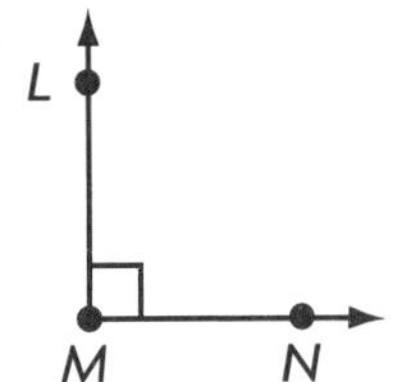

____

8. 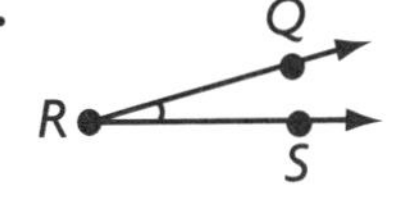

____

Mixed Applications

For Problems 9–10, use the road map.

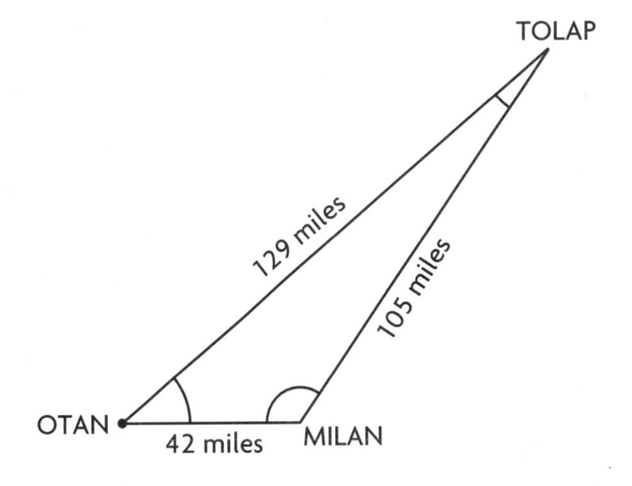

9. What is the measure of the angle formed by the roads with Otan as the vertex?

____

10. Sara went from Tolap to Milan to Otan and back to Tolap. How many miles did she travel?

____

Name ____________________

Problem-Solving Strategy

Draw a Diagram

Draw a diagram and solve.

1. Chad is planning to meet Sara at the Fruit Market. Sara gives Chad the directions to the Fruit Market. Make a map from Sara's directions.

Ride your bike south onto
Highway 17 for 5 miles. Make a
90° turn west on Highway 60.
Ride 2 miles. Make a 90° turn
south on Highway 11 and ride
1 mile. Look for the Fruit Market
on your right.

2. Henry is helping his father nail down a new roof. He puts a nail every 4 inches along one edge of the roof. He puts a nail at each end. The roof is 10 feet long. How many nails does Henry use?

____

3. Sam is decorating a cake. He plans to put a candy rose every inch around the edge. The cake is 20 inches around the edge. How many candy roses does he need?

____

Mixed Applications

Solve.

CHOOSE A STRATEGY

- **Write a Number Sentence**
- **Use a Schedule**
- **Make a Model**
- **Draw a Diagram**

4. On the way to the mountains, John, Mary, and Sue all rode in the back seat. John does not like to sit by a window. Mary does not like to sit behind the driver. From left to right, who sat where?

____

5. Fred bought some nails to build a bookcase. He bought $\frac{1}{4}$ pound of 1-inch nails, $\frac{1}{2}$ pound of $1\frac{1}{2}$-inch nails, and $\frac{1}{8}$ pound of 2-inch nails. How many pounds of nails did he buy?

____

6. Laura has $2.80 in change. She has pennies, nickels, and dimes. She has 6 nickels and 4 times as many dimes as nickels. How many pennies does she have?

____

7. Jan builds a rectangular fence 12 ft long and 8 ft wide. She places a post every 4 ft. How many posts does Jan need?

____

Name ______________________

Classifying Quadrilaterals

Vocabulary

Write the correct letter from Column 2.

Column 1

_____ 1. has 4 congruent sides and 2 pairs of congruent angles

_____ 2. has 2 pairs of congruent sides and 2 pairs of parallel sides

_____ 3. has 4 sides of any length and 4 angles of any size

_____ 4. has only 1 pair of parallel sides

Column 2

a. quadrilateral

b. trapezoid

c. parallelogram

d. rhombus

Draw and name the quadrilateral.

5. adjacent sides not equal; 2 pairs of congruent sides; 4 right angles

____

6. opposite sides not parallel; 4 sides and 4 angles

____

For Problems 7–10, write *true* or *false*.

7. A trapezoid has 4 right angles.

____

8. A square has 4 congruent sides.

____

9. A rectangle has 2 pairs of parallel sides.

____

10. A trapezoid has only one pair of parallel sides.

____

11. Dolores walked the perimeter of a park. It was 110 feet on all sides. Each corner was a 90° angle. What shape was the park?

____

12. Tom rode his bike for 3 hours every day after school for a week. His speed was 6 mph. How many miles did Tom ride his bike?

____

Name ______________________________

LESSON 23.5

Classifying Triangles

Vocabulary

Complete.

1. A triangle in which each side is a different length is ____________.

2. A triangle that has 2 congruent sides is ____________.

Name each triangle. Write *isosceles, scalene,* or *equilateral.*

3. 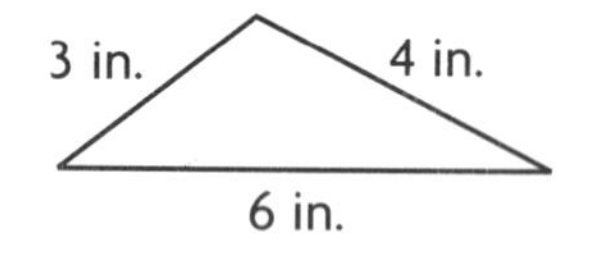

____

4. 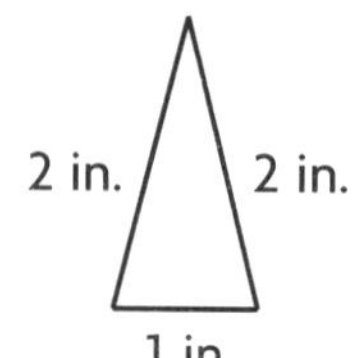

____

5. 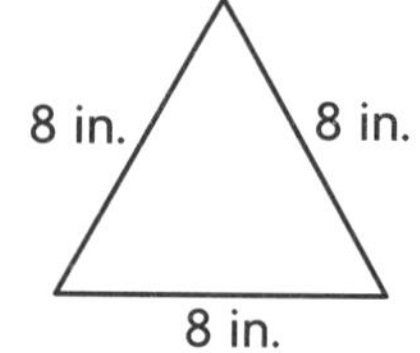

____

6. 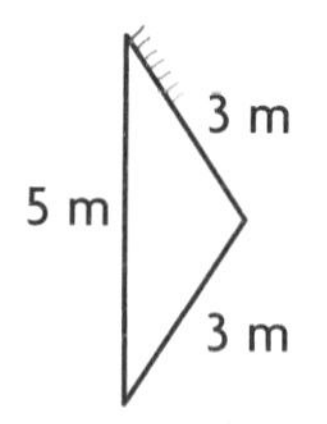

____

7. 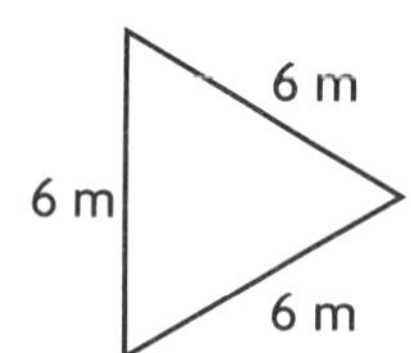

____

8. 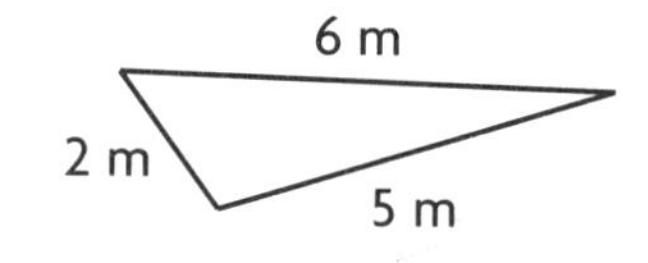

____

Measure the sides. Classify each triangle. Write *isosceles, scalene,* or *equilateral.*

9. 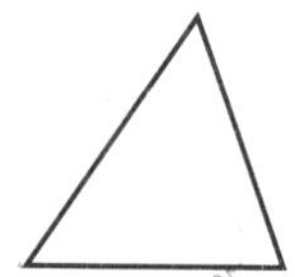

____

10. 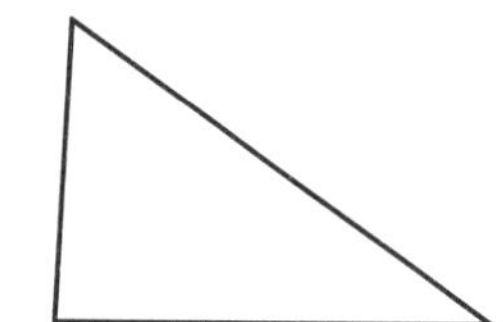

____

11. 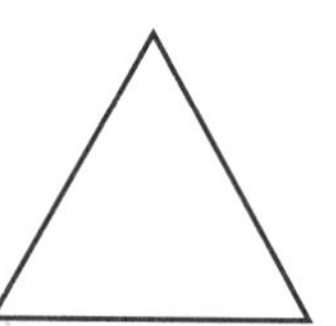

____

Mixed Applications

12. Julie wants to draw a triangle. She knows that one angle is 70° and another is 55°. How many degrees should the third angle be? How would you classify the triangle?

____

13. Larry was on a trip with his father. They drove for 4 hours at 60 mph. After lunch they drove $3\frac{1}{2}$ hours at 40 mph. How far did they drive in all?

____

Name ______________________________

More About Classifying Triangles

Vocabulary

Write the correct letter from Column 2.

Column 1	Column 2
______ 1. a triangle that has one obtuse angle	a. right triangle
______ 2. a triangle that has three acute angles	b. acute triangle
______ 3. a triangle that has a right angle	c. obtuse triangle

Name each triangle. Write *right*, *acute*, or *obtuse*.

4. 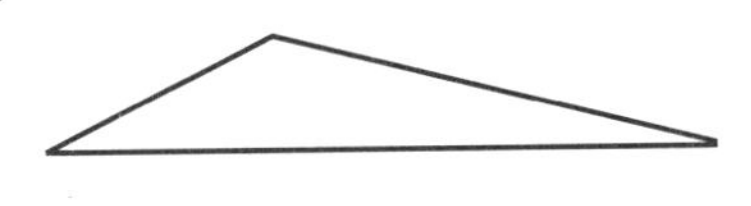

____

5. 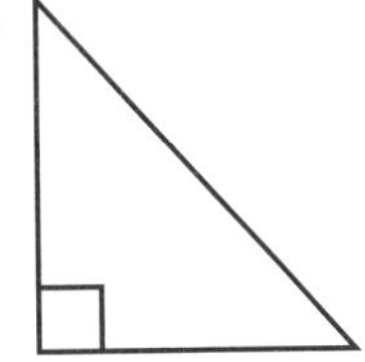

____

6. 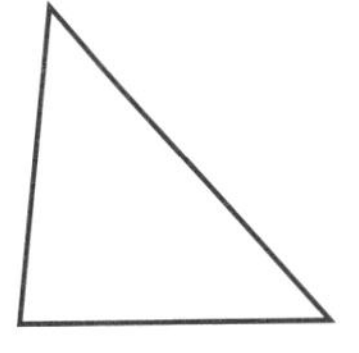

____

Find the measure of the unknown angle in each triangle.

7. 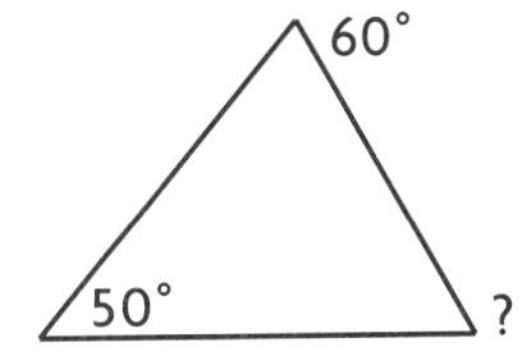

____

8. 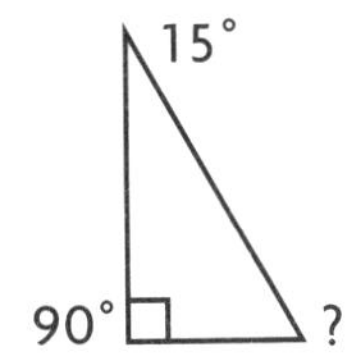

____

9. 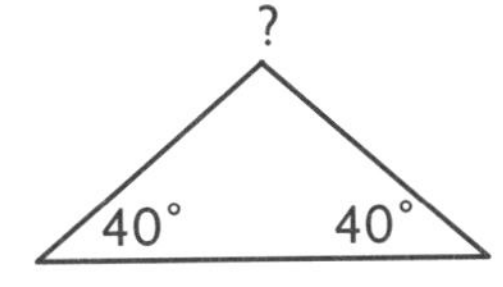

____

Mixed Applications

10. Kyle cut a rectangular-shaped cake in half by cutting it from corner to corner. He made two triangular-shaped cakes. How would you classify the triangles he formed?

____

11. Patty started her test at 8:30. She finished the test in 1 hour 20 minutes. Then she spent 1 hour 35 minutes in the library. At what time was she done?

____

Name ____________________

Testing for Congruence

Use a ruler and a protractor to test each pair of figures for congruency. Write *congruent* or *not congruent*.

1.

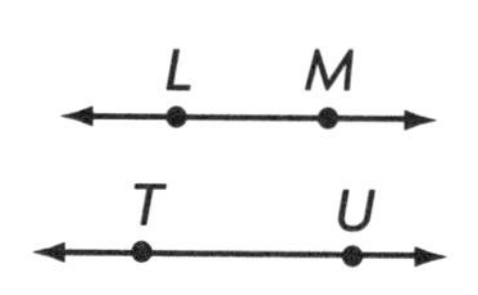

____

2.

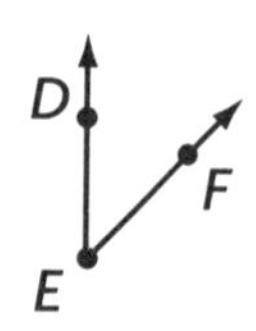

____

3.

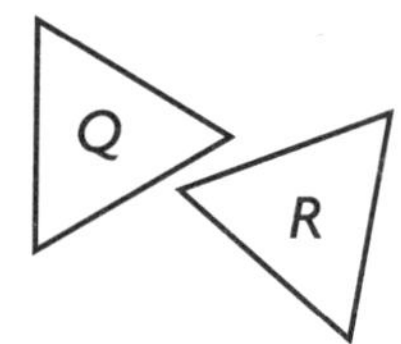

____

Write the letters of the two figures that are congruent.

4. a.

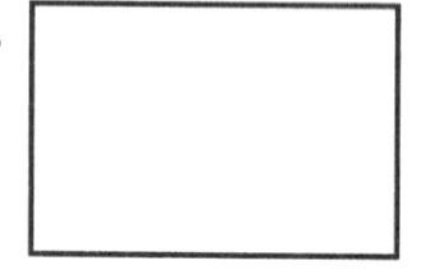

b.

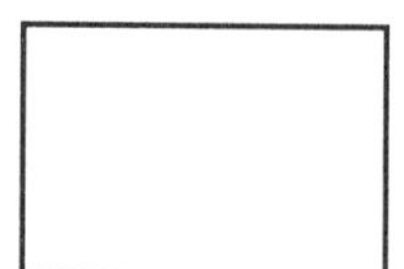

c.

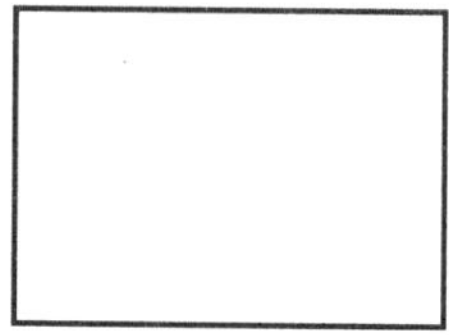

____

5. a.

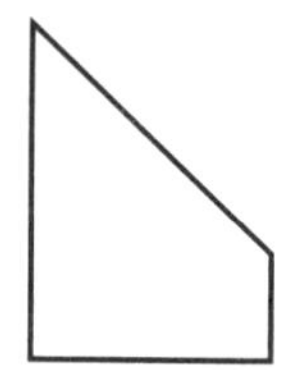

b.

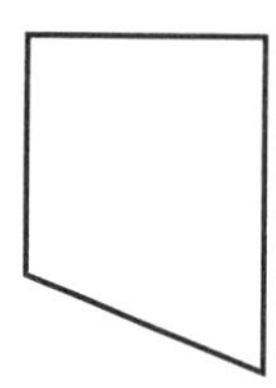

c.

____

Mixed Applications

For Problem 6, use the drawing.

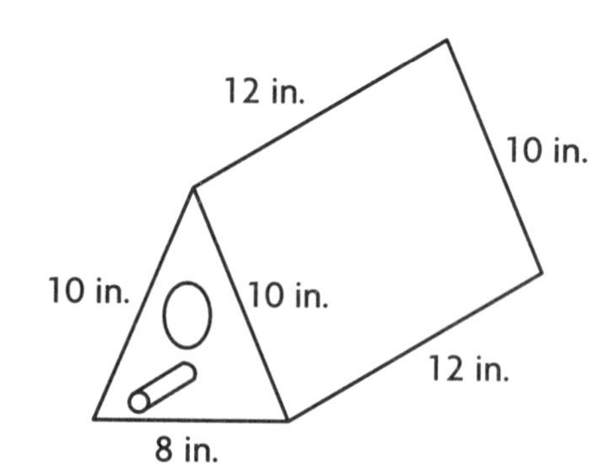

6. Grace and Harry are building birdhouses to raise money for charity. Including the end, side, and bottom that are not seen, they use 5 pieces of wood and a round dowel. Name the pieces that are congruent.

____

7. They plan to sell birdhouses for $8.50 each. It costs $1.45 for materials to make one. If they sell 10 birdhouses how much money will they make?

____

Name ________________________________

LESSON 24.2

Congruence and Symmetry

Tell whether the two halves of each drawing are congruent. Write *yes* or *no*.

1. 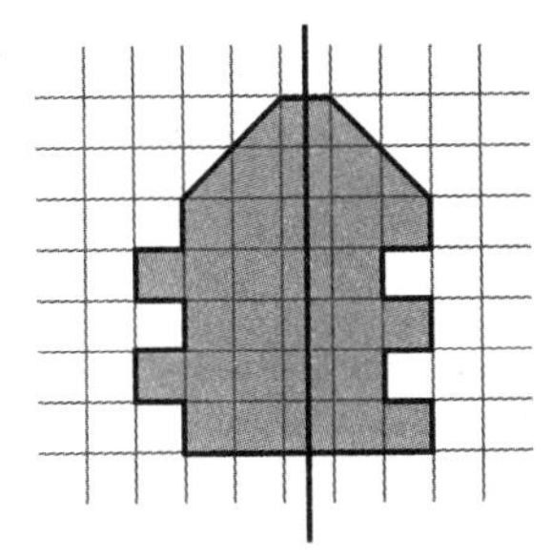

2. 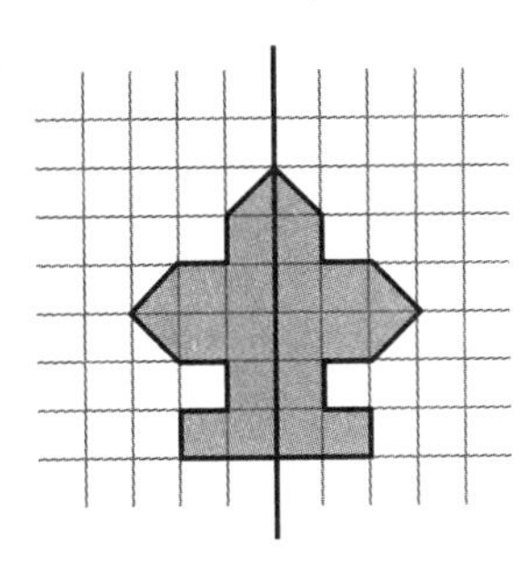

3. 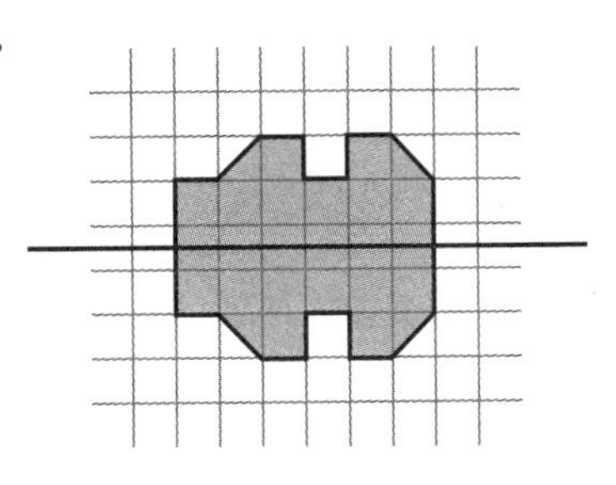

______________ ______________ ______________

Draw the lines of symmetry for each figure.

4. 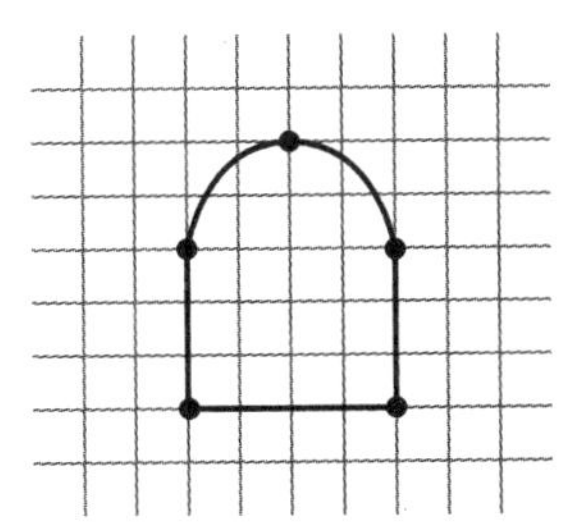

5. 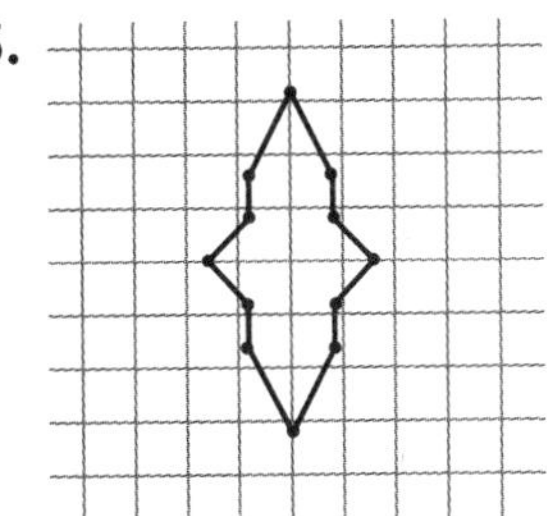

6. 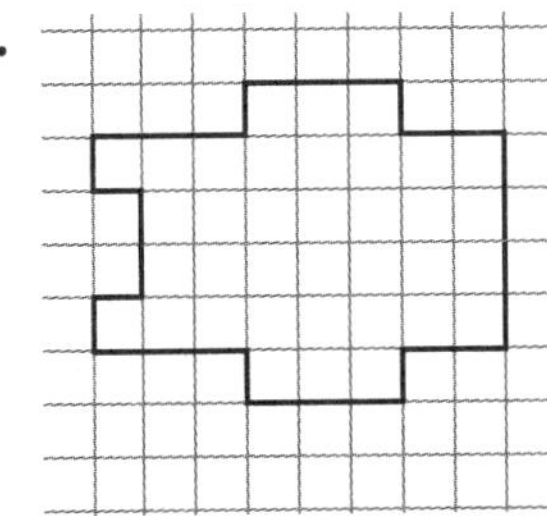

7. 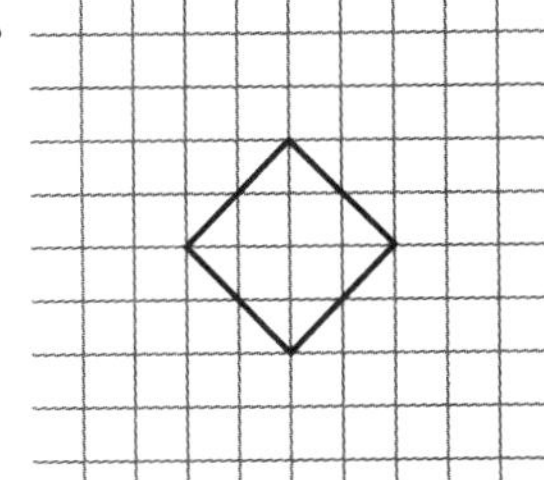

8. 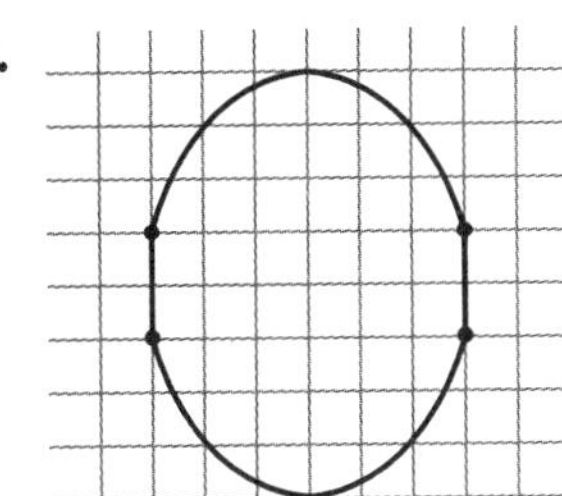

9. 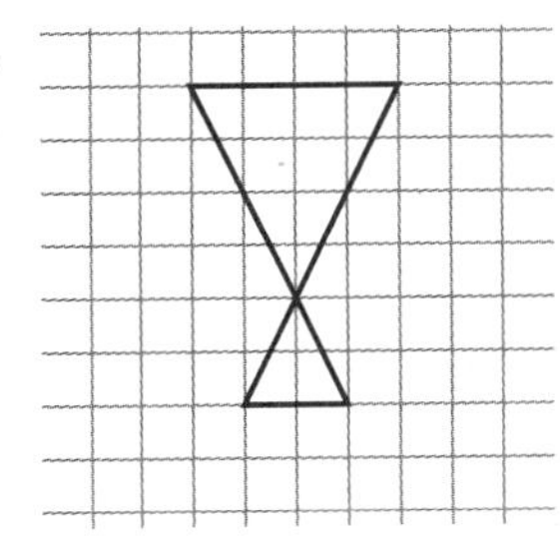

Mixed Applications

For Problems 10–11, use the letters.

10. Which letter has both point and line symmetry?

____

H  T

11. Which letter has exactly two lines of symmetry?

____

12. Find the change from a $20 bill for purchases totalling $17.21.

____

Name ______________________________

LESSON 24.3

Transformations on the Coordinate Grid

Vocabulary

Complete.

1. When you move a figure to show a translation, reflection, or rotation, it is called a ______________________.

Translate, reflect, and rotate each figure on the coordinate grid. Draw the new figure with its coordinates. Name the new ordered pairs.

2.

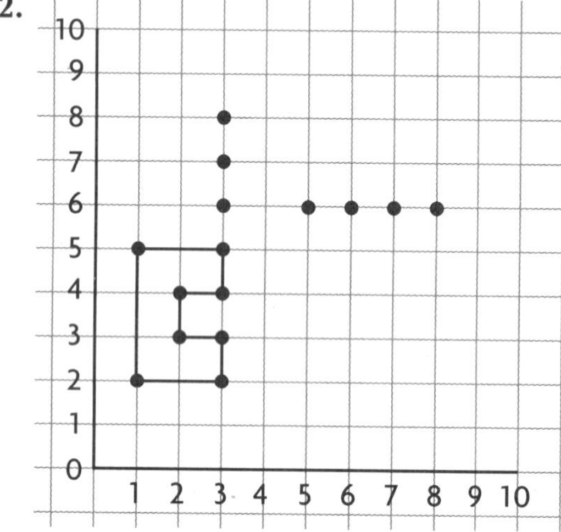

3.

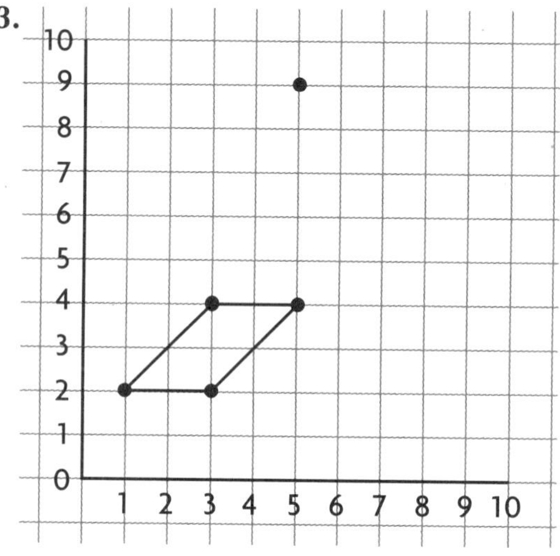

Mixed Applications

4. Sid wants to move his doghouse from its location at (2,3), (4,3), (2,6), and (4,6). The new ordered pairs for the location are (5,1), (5,3), (8,1), and (8,3). Use a coordinate grid to draw the figure representing the doghouse in its new location. Explain how the doghouse was moved.

____

____

____

5. Wesley started walking at 9:30 A.M. He took a 45-minute break for lunch and two 10-minute breaks to rest. He finished his walk at 4:50 P.M. How long was Wesley walking?

____

Name ______________________________

LESSON 24.4

Tessellations

Vocabulary

Complete.

1. When closed figures are arranged to cover a surface with no gaps and no overlaps, it is called a ______________________.

Trace and cut out each figure. Write *yes* or *no* to tell whether each figure tessellates.

2. 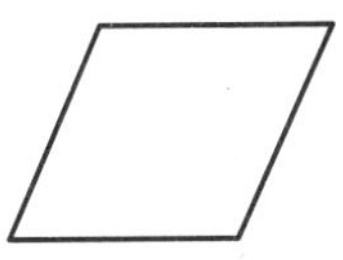

____

3. 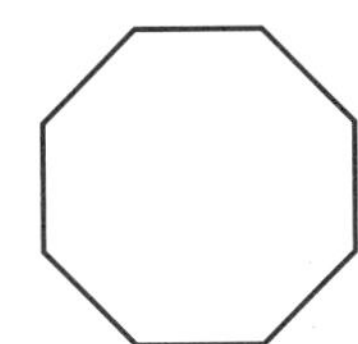

____

4. 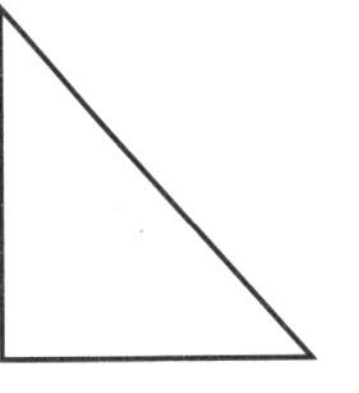

____

Trace and cut out each figure. Translate, reflect, or rotate it to make a design that tessellates.

5. 

6. 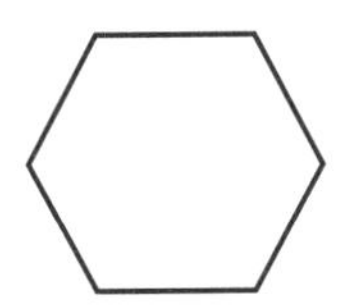

7. 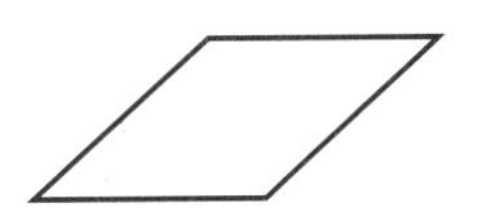

8. 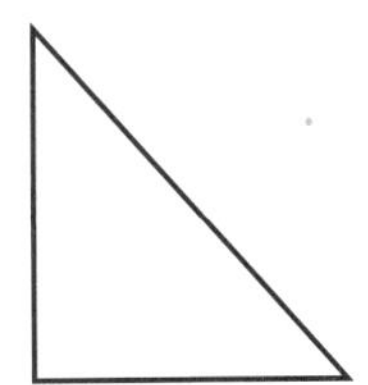

Mixed Applications

9. Shirley is buying 12-inch-square tiles for $1.55 each. She wants to cover her bathroom floor. It is 9 ft long and 9 ft wide. How many tiles will she need? What will they cost?

____

10. Dennis is using hexagonal and triangular tiles to cover his wall. Draw the design that Dennis will have on his wall.

Name ______________________________

LESSON 24.4

Problem-Solving Strategy

Make a Model

Make a model to solve.

1. Fred is making a tabletop with a mosaic design. He can use two or more polygons. The polygons need to tessellate to cover the tabletop. Use pattern blocks to make a design that tessellates. Draw your design.

2. Art has pattern blocks that are a small equilateral triangle, small square, and small parallelogram. Which two of these blocks will tessellate when arranged in a design? Draw the design.

3. Betty wanted to budget her free time on weekends. She planned to spend $\frac{1}{2}$ of her time on chores, $\frac{1}{3}$ of her time playing with friends, and $\frac{1}{6}$ of her time reading. She read for 2 hours. How many hours did she play with her friends?

____

4. Tony makes picture frames from wood molding to sell at the craft fair. He has 40 inches of molding. He wants to create the largest possible picture frame. What dimensions should the frame be to give the greatest possible area?

____

Mixed Applications

Solve.

CHOOSE A STRATEGY

- Make a Graph • Write a Number Sentence • Work Backward • Make a Model • Draw a Diagram

5. Marcus is making a design using hexagons. Above each hexagon, he wants to put two trapezoids. His design calls for 215 hexagons. How many trapezoids will he need for his design?

____

6. Jane wants to tile her 8 ft by 10 ft dining area and her 12 ft by 12 ft kitchen with the same tile. How many square feet does she have to cover?

____

Name ______________________________

Construct a Circle

Vocabulary

Write the correct letter from Column 2.

Column 1		Column 2
1. chord	____	a. A tool for constructing circles
2. diameter	____	b. A line segment that connects the center with a point on the circle
3. circle	____	c. A line segment that connects any two points on the circle
4. radius	____	d. A closed figure with all points on the figure the same distance from the center point. It has no beginning point and no ending point.
5. compass	____	e. A chord that passes through the center of the circle

Use a compass to construct the circle and find the measurements.

6. Construct a circle with a radius of 2 cm. Label and measure a diameter, radius, and chord.

Write *chord, diameter,* or *radius* for each line segment.

7. 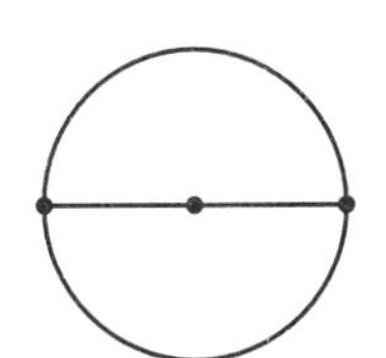

____

8. 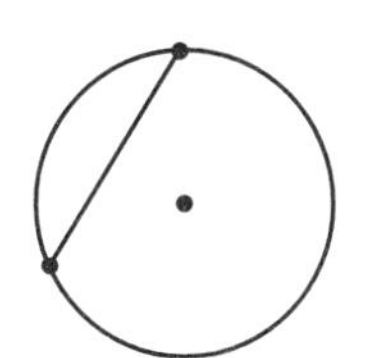

____

9. 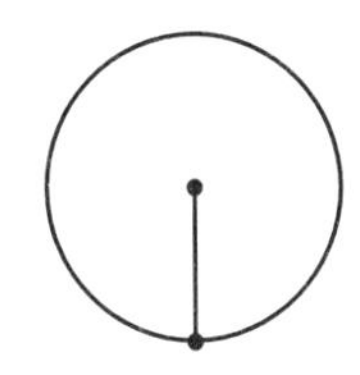

____

Mixed Applications

10. Josephine knows that the diameters of the two hoops are 20 in. and 18 in. What is the radius of each hoop?

____

11. Alice was in the library from 3:30 to 5:45 each afternoon after school this week. How long did she spend in the library?

____

Name ____________________

LESSON 25.2

Finding Circumference

Vocabulary

Fill in the blank.

1. The perimeter of a circle is the ____________________.

2. Use a ruler and string to find the circumference of a pencil or crayon.

Use a calculator to divide the circumference of each object by its diameter. Complete the table.

Object	Circumference (c)	Diameter (d)	$\frac{c}{d}$
3. Plate	12.6 in.	4 in.	________
4. Tire	125.6 in.	40 in.	________
5. Jar	15.7 in.	5 in.	________

Mixed Applications

For Problems 6–7, use the pictures.

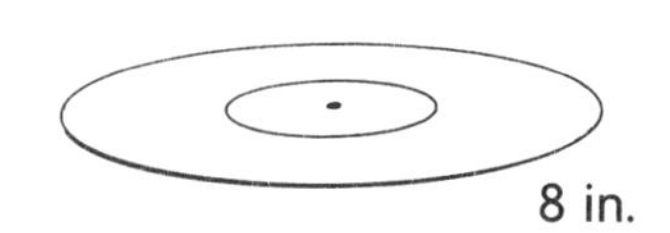

6. What is the perimeter of the smallest box that this record can fit into?

____

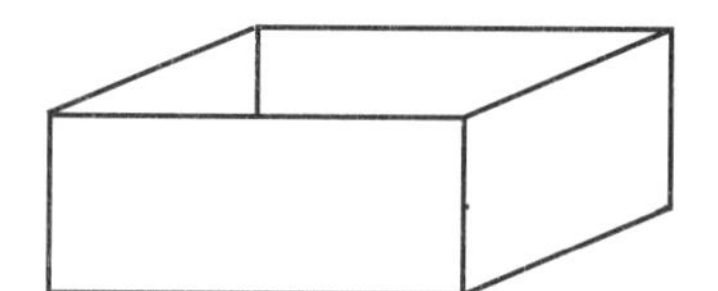

7. What is the radius of the record?

____

8. A record costs $4.99. Joe wants to buy one for himself and one for each of his 5 friends. How much would it cost Joe for the records?

____

9. Florence has 56 stickers. She wants to give 4 friends her stickers. How many stickers will each friend get?

____

Name ______________________________

Problem-Solving Strategy

Act It Out

Act it out to solve. Round to the nearest tenth when necessary.

1. Sylvia is making a flower vase. She wants to cover the cylindrical part of the vase with tissue paper. The cylinder is 18 cm in height, and its diameter is 7 cm. What is the circumference of the cylinder? What size should Sylvia cut the tissue paper to fit the cylinder exactly?

____

____

2. Robert made a cylinder out of clay. He wants to cover it with a piece of newspaper. The cylinder is 14 in. high and has a diameter of 5 in. The piece of newspaper he has is 18 in. by 16 in. Does Robert have enough paper? Explain.

____

____

3. Melanie bought a book for $12.75. She handed the clerk a $20 bill. What combination of bills and coins could she get back as her change?

____

____

4. Tyler had two $5 bills, two $10 bills, one $20 bill, one $1 bill, 2 quarters, 2 dimes, 1 nickel, and 4 pennies. He spent $12.00. He put $29.41 in the bank. What bills and coins does Tyler have left?

____

____

Mixed Applications

Solve.

CHOOSE A STRATEGY

- Work Backward
- Make a Model
- Draw a Diagram
- Act It Out
- Write a Number Sentence

5. Fiona needs to wrap a cylindrical box. It is 14 in. in height, and its diameter is 8 in. What is the circumference of the cylinder? What size paper will Fiona need?

____

6. Derek, Rose, Juan, Laura, and Jeff are in line at the snack counter. Jeff is in front of Rose and after Derek. Juan is between Laura and Derek. Who is last in line?

____

Name ______________________________

LESSON 25.3

Angles in a Circle

Write *true* or *false* to describe each statement.

1. There are four 90° angles in a circle.

   ____

2. There are 361° in a circle.

   ____

3. You can divide a right angle into two acute angles.

   ____

4. If six of seven angles in a circle measure 55°, the seventh angle measures 55° too.

   ____

Find the missing angle.

5. 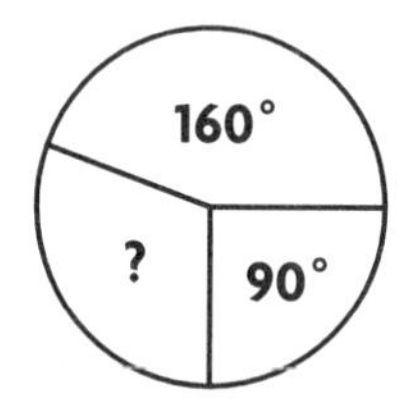

   ____

6. 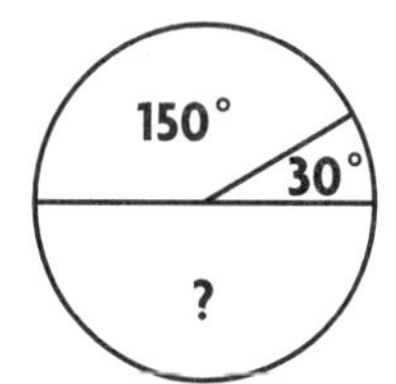

   ____

7. 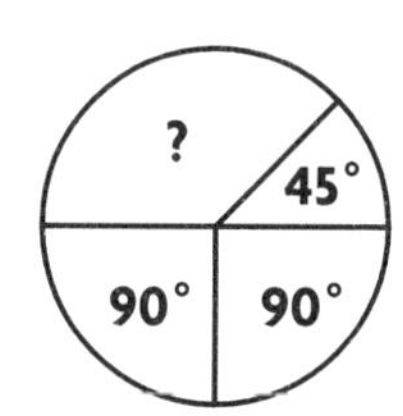

   ____

8. 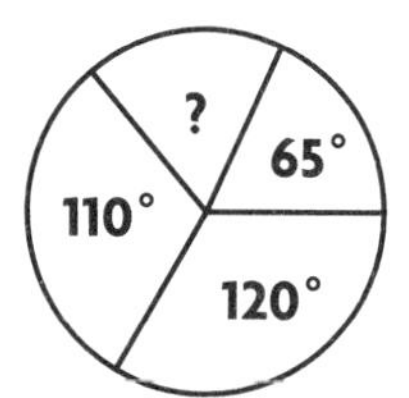

   ____

Mixed Applications

For Problem 9, use the clocks.

9. Which clock shows an obtuse angle?

   ____

Clock A

Clock B

10. Toni has a jar she wants to decorate with felt. The jar is 15 cm high. Its diameter is 10 cm. What is the circumference of the jar? What size felt will Toni need?

    ____

11. Jon has 4 tickets to a soccer game. If each ticket costs $5.95, how much did Jon spend on tickets?

    ____

Name ______________________________

LESSON 25.4

Measuring Angles in a Circle

Use a protractor to find the number of degrees in the angles of the circle.

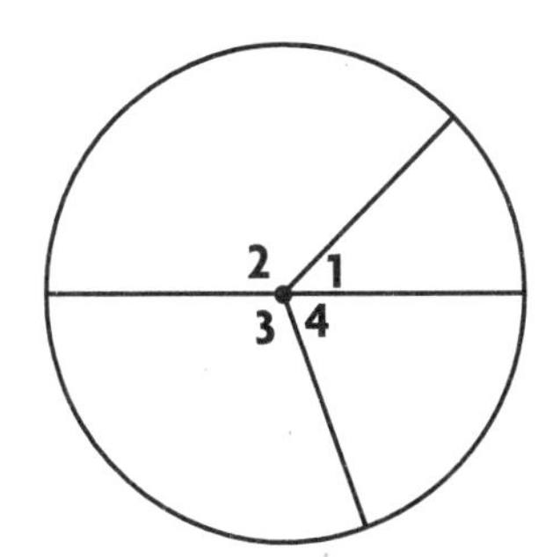

1. How many degrees are in angle 4?

2. How many degrees are in angle 3?

____

3. How many degrees are in angle 1?

4. How many degrees are in angle 2?

____

5. How many degrees are in angles 1, 2, 3, and 4 in all?

____

Use a compass and a protractor to draw a circle with the following angles.

6. 5 angles of 72° each

7. 4 angles of 90° each

8. 2 angles of 130° each and 2 angles of 50° each

Mixed Applications

9. The school bus will pick up students only within a 2-mile radius of the school. To get to Lisa's house from school, you drive 2 miles south, make a 90° turn and drive 2 miles west. Can Lisa ride the school bus? Why or why not?

____

____

10. Chad went to the store. He bought 2 T-shirts for $12 each, a pair of jeans for $25, and 3 pairs of shoes for $21 each. How much money did Chad spend in all?

____

Name ______________________________

LESSON 26.1

Prisms and Pyramids

Vocabulary

Fill in the blanks.

1. A ________________ is a solid figure that has two congruent faces called ________________.

2. A ________________ is a solid figure with one ________________ that is a polygon and three or more faces that are triangles with a common vertex.

Write *prism* or *pyramid*. Write the polygon that names the base. Identify the solid figure.

3.

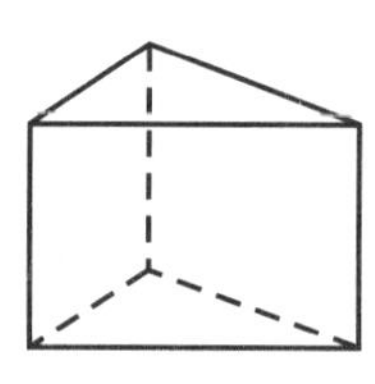

____

____

4.

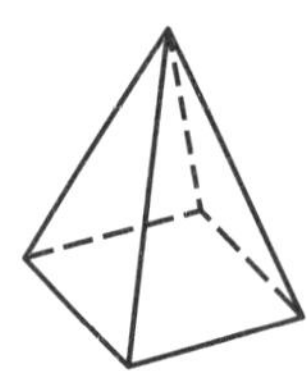

____

____

5.

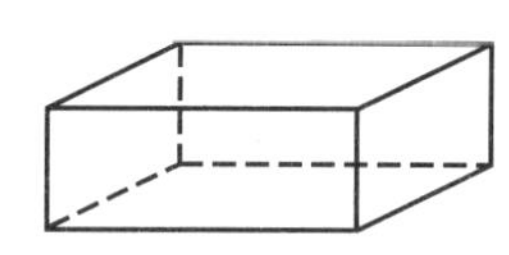

____

____

Write the name of the solid figure. Make a drawing of each.

6. I have 2 congruent pentagons for bases. I have 5 rectangular faces.

____

7. I have a base with 8 equal sides. My faces are 8 triangles.

____

Mixed Applications

8. Name two items that look like a cube.

____

____

9. Scott spent 18 hours driving to college. If his average speed was 55 mph, how many miles did Scott drive?

____

Name ______________________________

Nets for Solid Figures

Vocabulary

Complete.

A ______________ is a two-dimensional pattern for a three-dimensional solid.

Match each solid figure with its net. Write *a, b, c,* or *d*.

1. 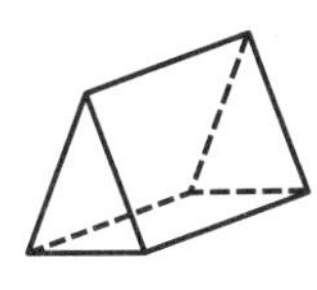 ________

2. 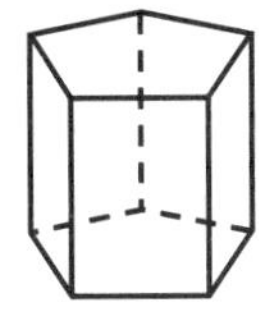 ________

3. 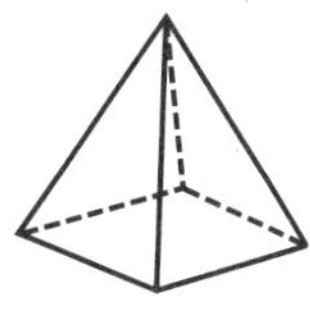 ________

4. 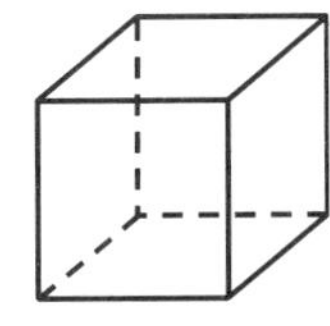 ________

a. 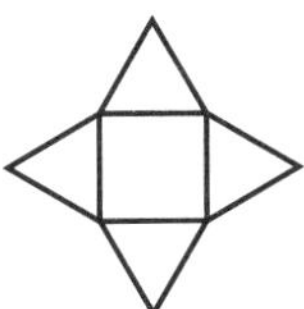

b. 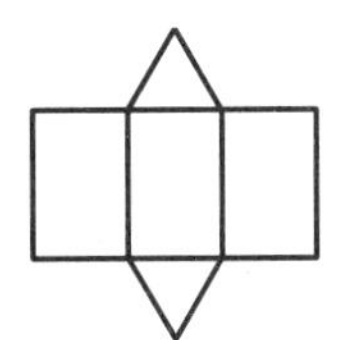

c. 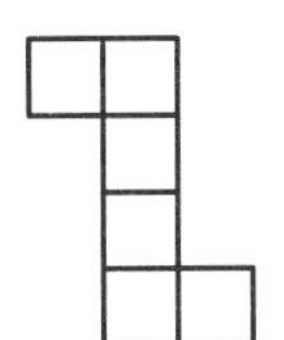

d. 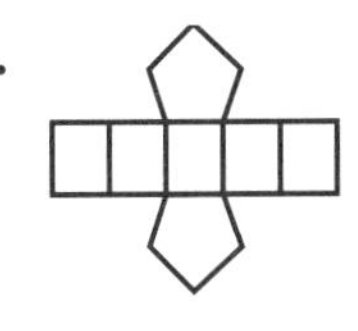

Circle the letter of the pattern that can be folded to make the figure.

5. 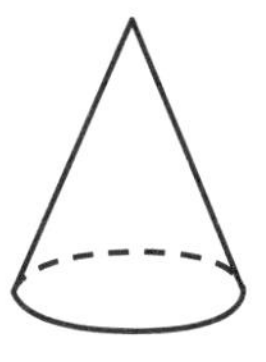

a. 

b. 

c. 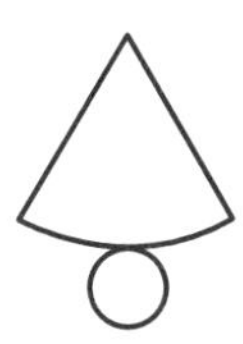

6. 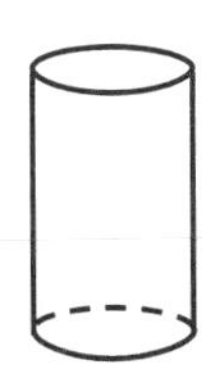

a. 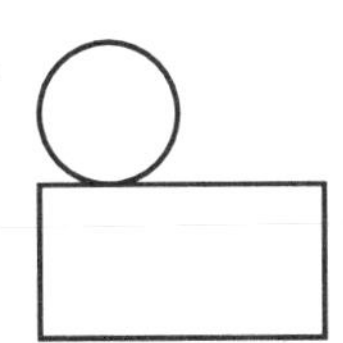

b. 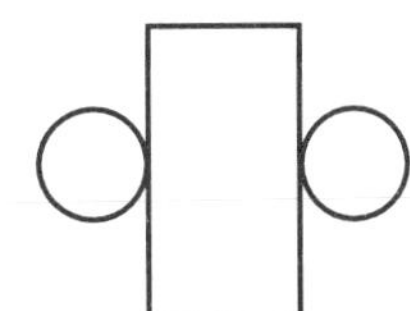

c. 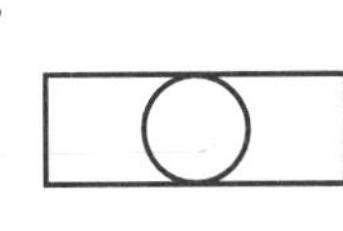

Mixed Applications

7. What faces would you find in a net for a square pyramid?

____

____

8. Cara earns $36.75 a week for 7 hours of babysitting. How much does she earn in 4 weeks? How much does she earn an hour?

____

____

Name ______________________________

LESSON 26.3

Solid Figures from Different Views

Use grid paper to draw each figure from the top, the side, and the front.

1. 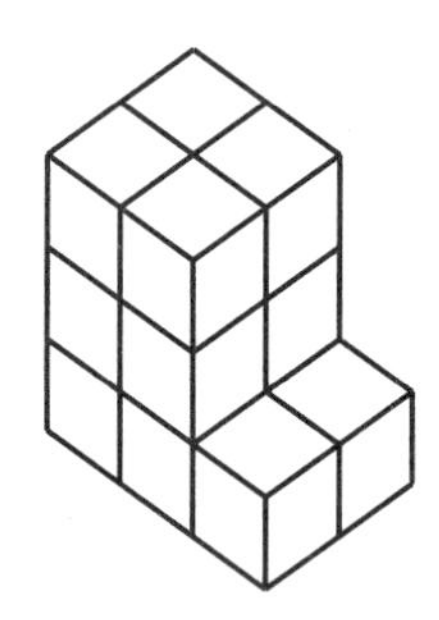

2. 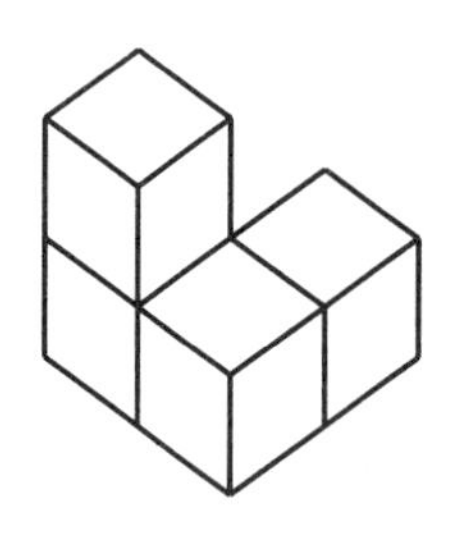

3. 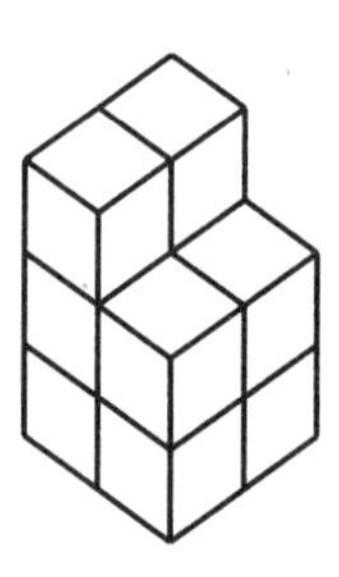

Choose the figure that is represented by each set of three drawings.

From the top | From the side | From the front

4. 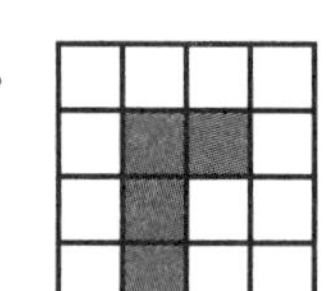 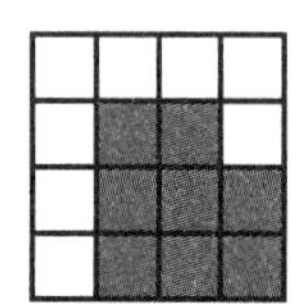 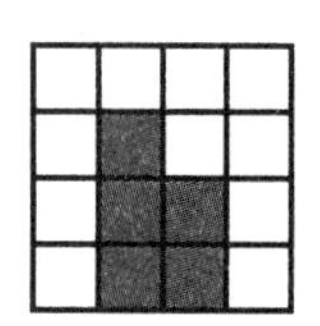 ____

a. 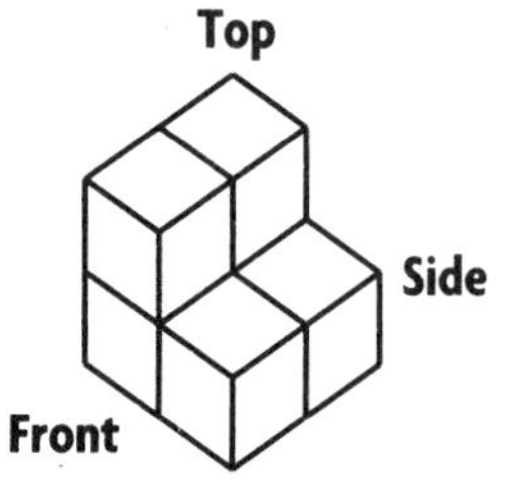

5. 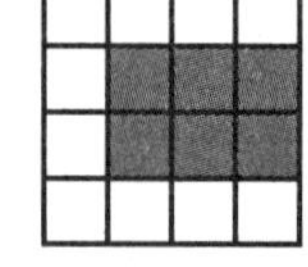 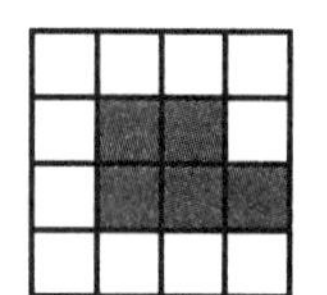 ____

b. 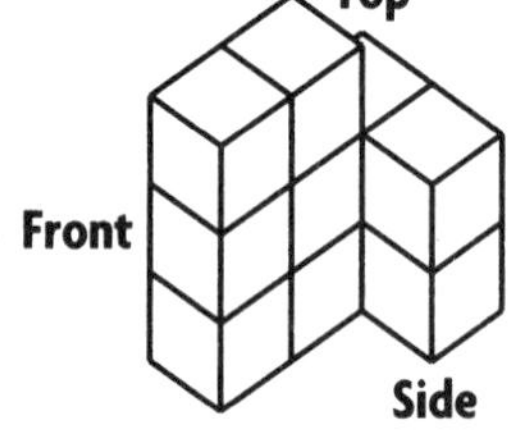

6. 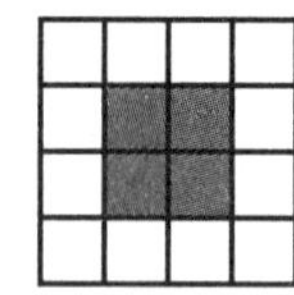 ____

c. 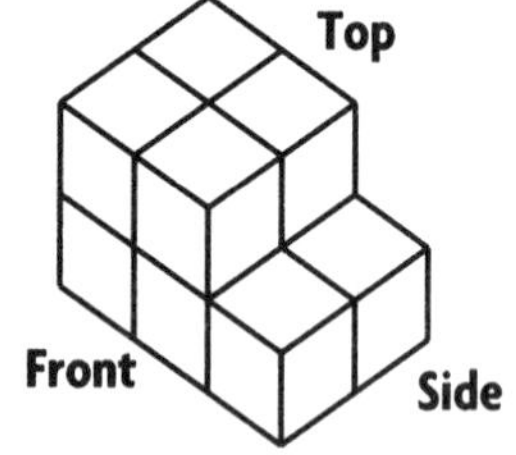

Mixed Applications

7. Bricks are stacked in 8 rows with 6 blocks in each row. If 10 layers of bricks are stacked, how many are in the stack?

____

8. A survey of 90 drivers found that $\frac{5}{6}$ of the people had taken driving lessons. How many of the drivers took driving lessons?

____

Name ____________________

LESSON 26.4

Algebraic Thinking: Volume

Find the missing dimension. You may use a calculator.

1.

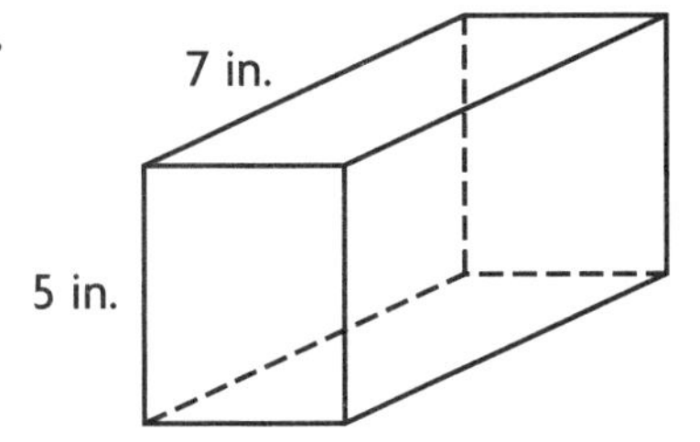

Volume = 140 cu in.

width = ________

2.

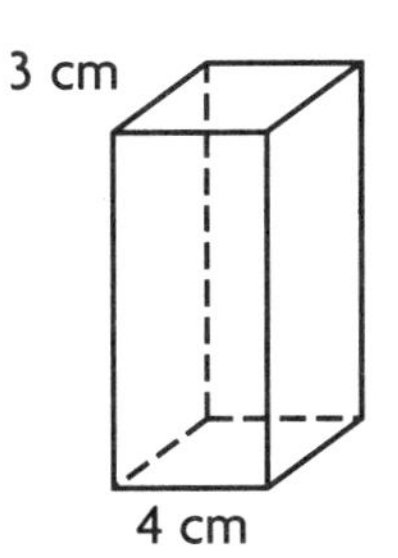

Volume = 108 cu cm

height = ________

3.

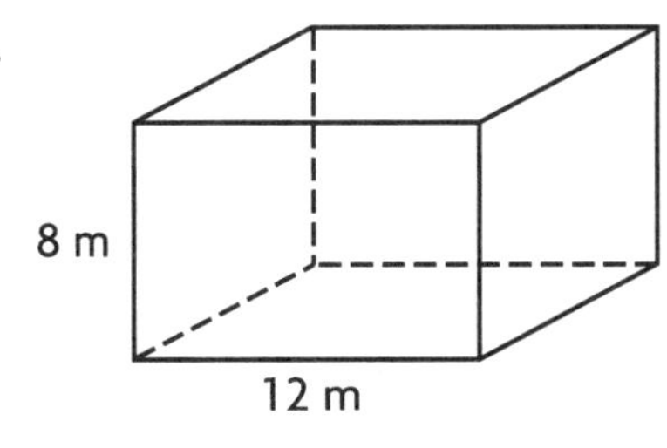

Volume = 672 cu m

length = ________

4. length = 11 yd

width = 5 yd

height = ________

Volume = 165 cu yd

5. length = 14 ft

width = 9 ft

height = 4 ft

Volume = ________

6. length = 8 in.

width = ________

height = 9 in.

Volume = 288 cu in.

Complete the table. You may use a calculator.

	Length	Width	Height	Volume
7.	15 in.	________	2 in.	240 cu in.
8.	________	11 m	5 m	385 cu m
9.	6 yd	8 yd	________	288 cu yd
10.	5 cm	3 cm	15 cm	________

Mixed Applications

11. A carton is 9 inches wide, 6 inches long, and 5 inches high. What is the volume of the carton?

____

12. Tom wants to buy a stereo that costs $540. He has saved $\frac{1}{3}$ of the cost. How much has Tom saved?

____

Name ______________________________

LESSON 26.4

Problem-Solving Strategy

Use a Formula

Use a formula and solve.

1. A garden that is 18 feet wide and 22 feet long needs to be fenced. Will 25 yards of fencing be enough? Explain.

____

____

2. The trailer of a lumber truck is 15 feet wide, 18 feet long, and 10 feet high. Is the truck large enough to carry 2,500 cubic feet of lumber?

____

____

3. Tim is packing a box that is 18 inches long and 12 inches wide and has a volume of 3,240 cubic inches. He wants to pack an object that is 9 inches long, 6 inches wide, and 16 inches high. Will the object fit in the box? Explain.

____

____

4. New flooring is being installed in the school foyer. The area is 15 feet wide and 33 feet long. How many square yards of flooring is needed? What is the perimeter of the foyer, measured in feet?

____

____

Mixed Applications

Solve.

CHOOSE A STRATEGY

- **Draw a Diagram**
- **Make a Table**
- **Use a Formula**

5. Classes at the high school begin at 7:45 A.M. Each class is 50 minutes long, and there is a 7-minute break after each class. At what time does the second class of the day end?

____

6. A swimming pool is 60 feet long and 30 feet wide. How many cubic feet of water will be needed to fill the pool to a depth of 8 feet?

____

Name ____________________

Estimating Volume

Use the benchmarks at the right to name the more reasonable unit for measuring the volume of each box. Estimate the volume.

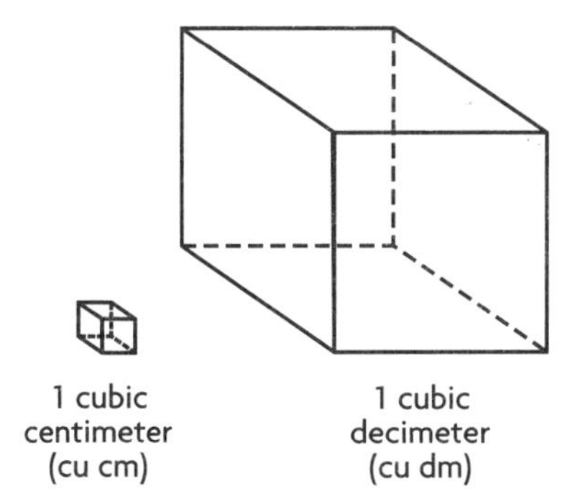

1.

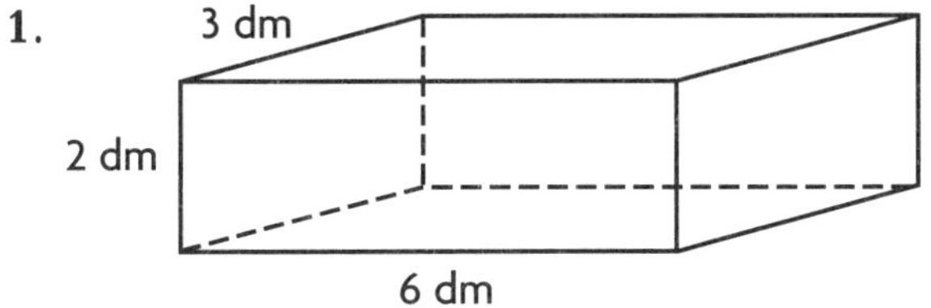

2.

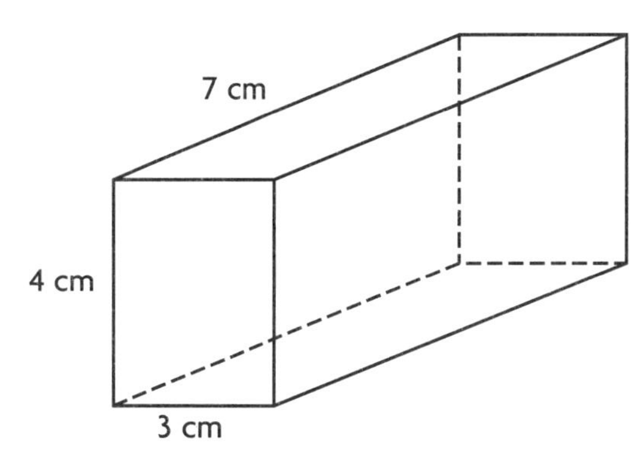

Choose the most reasonable measure. Circle *a, b,* or *c.*

3. a desk a. 25 cu in. b. 25 cu ft c. 25 cu yd

4. a breadbox a. 360 cu in. b. 360 cu ft c. 360 cu yd

5. a barn a. 480 cu in. b. 480 cu ft c. 480 cu yd

Mixed Applications

6. A carton is 12 inches long and 15 inches wide. Jon estimates the volume of the carton as 2,000 cubic inches. Ted says the estimated volume is 200 cubic inches. Which estimate is more reasonable?

7. A number is greater than 42 but less than 50. The sum of its digits is 11. What is the product of that number and 16?

Name ______________________________

LESSON 27.1

Understanding Ratio

Vocabulary

Fill in the blank.

1. A ____________ can be used to compare two numbers in three ways.

Name the type of ratio.

2. There were 4 baseballs and 6 basketballs.

   ____

3. Margo had 3 quarters and 2 pennies.

   ____

4. Recess is preferred by 19 of 20 students.

   ____

5. Of 20 students, 11 are boys.

   ____

Use the picture to make the comparison.

6. 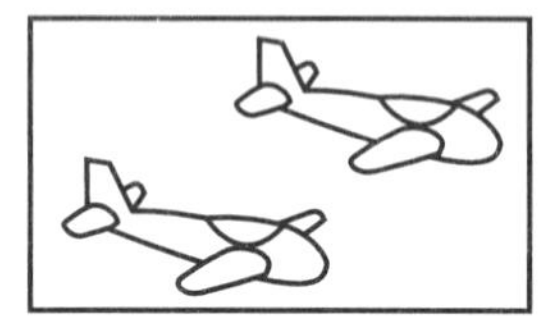

   _____ wings to _____ planes

7. 

   _____ flowers to _____ stem

8. 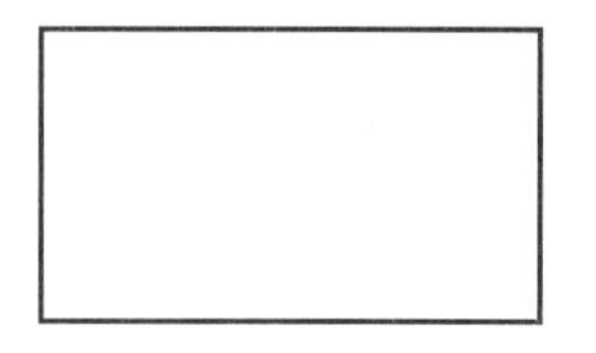

   _____ legs to _____ spiders

9. 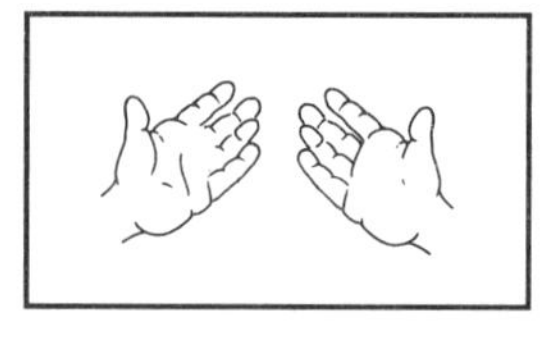

   _____ fingers to _____ hands

Mixed Applications

10. Last month there were 21 days of school and 10 days off. What was the ratio of school days to total days?

    ____

    ____

11. Mavis ran the 200-meter dash in 25.43 seconds. What does the digit 4 stand for?

    ____

    ____

Name ______________________________

Expressing Ratios

Write *a* or *b* to show which fraction represents the ratio.

1. 7 to 9 ______

a. $\frac{9}{7}$ b. $\frac{7}{9}$

2. 6:2 ______

a. $\frac{6}{2}$ b. $\frac{2}{6}$

3. 9:3 ______

a. $\frac{9}{3}$ b. $\frac{3}{9}$

4. 11 to 16 ______

a. $\frac{16}{11}$ b. $\frac{11}{16}$

For Exercises 5–7, use the table. Write each ratio in three ways.

5. What is the ratio of race games to sports games?

____

6. What is the ratio of all games to arcade games?

____

7. What is the ratio of sports games to all games?

____

BEN'S VIDEOGAME COLLECTION	
Number of Games	**Type**
5	Race
3	Arcade
2	Sports

Mixed Applications

8. Becky took a survey of her classmates' pets. The students had 9 dogs and 13 cats. What was the ratio of cats to dogs?

____

9. Kareem found that one-sixth of his jelly beans were red. If he had 84 jelly beans, what was the ratio of red jelly beans to the whole?

____

10. Brian had a triangle with a 90° angle and a 19° angle. How many degrees were in the third angle?

____

11. Erik discovered he was $\frac{3}{4}$ as tall as Wilt Chamberlain, the NBA legend. Chamberlain is 86 inches tall. How tall is Erik?

____

Name ______________________

LESSON 27.3

Equivalent Ratios

Vocabulary

Fill in the blank.

1. ______________________ are ratios that show the same relationship.

Tell whether the ratios are equivalent. Write *yes* or *no*.

2. $\frac{3}{8}$ and $\frac{9}{24}$ __________

3. 4:5 and 5:4 __________

4. 7 to 4 and 28 to 16 __________

5. $\frac{8}{4}$ and $\frac{2}{1}$ __________

6. 6:8 and 2:4 __________

7. 3 to 15 and 4 to 20 __________

8. Complete the table.

Number of oranges to make orange juice	5	____	____	____
Pints of orange juice	1	2	3	4

Write three ratios that are equivalent to the given ratio.

9. 7:1 ______________________

10. 6:3 ______________________

11. 3 to 2 ______________________

____

12. 13 to 15 ______________________

____

13. $\frac{5}{2}$ ______________________

14. $\frac{7}{8}$ ______________________

Mixed Applications

15. There are 35 campers at Backbush summer camp. Of those, 28 play baseball. What is the ratio of baseball players to all campers? Write the ratio in simplest form.

____

16. Katie is buying orange juice. She can buy 1 gallon for $2.75 or 1 pint for $0.35. Which is the better buy?

____

Name ______________________________

LESSON 27.4

More About Equivalent Ratios

Vocabulary

Fill in the blank.

1. A ratio that compares the distance on a map with the actual distance is a ______________.

Copy and complete each ratio table.

2.	Gallons of gas used	1	2	____	7	____
3.	Miles traveled	18	36	90	____	198

4.	Number of minivans	1	4	7	____	15
5.	Number of passengers	7	28	____	84	____

For Problems 6–9, use the drawing of the patio.

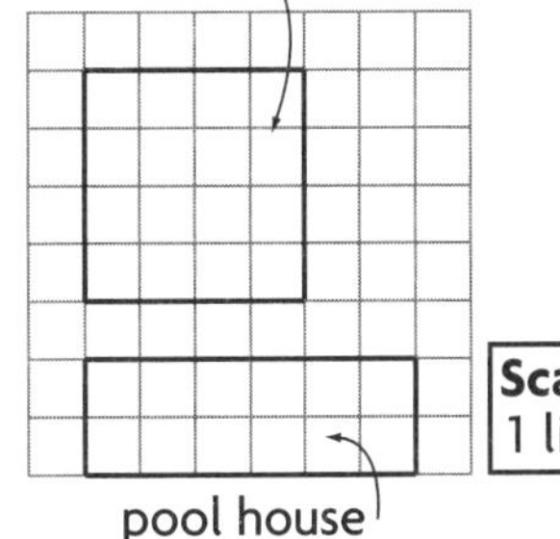

6. What is the width of the pool in units?

____

7. What is the actual width of the pool?

____

8. What is the perimeter of the pool house in units? in feet?

____

9. What is the ratio of linear units to feet?

____

10. A builder plans to use a scale of 2 cm = 5 ft in her plan for a house. What length will the lines be that represent 5 ft, 10 ft, 15 ft, 20 ft, and 25 ft?

____

11. A yard is 28 ft long and 36 ft wide. It costs $0.50 per square foot to have grass planted. What is the total cost?

____

Name ______________________________

LESSON 27.5

Ratios in Similar Figures

Vocabulary

Fill in the blank.

1. In ____________________ figures, the matching angles are congruent and the sides have equivalent ratios.

Write *yes* or *no* to tell whether the shapes are similar.

2.

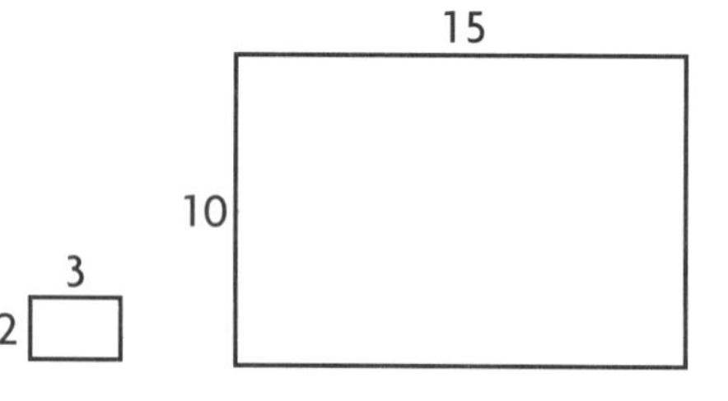

____

3.

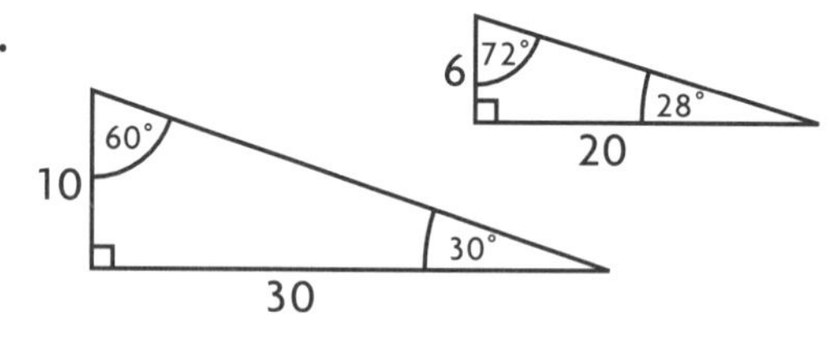

____

Find the length of the missing side in the similar shapes.

4.

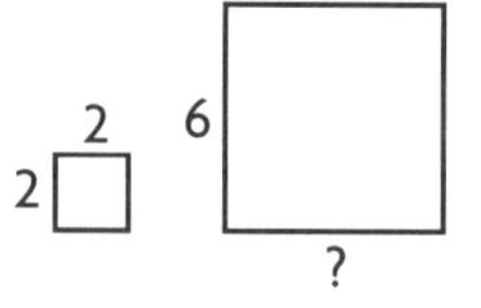

____

5.

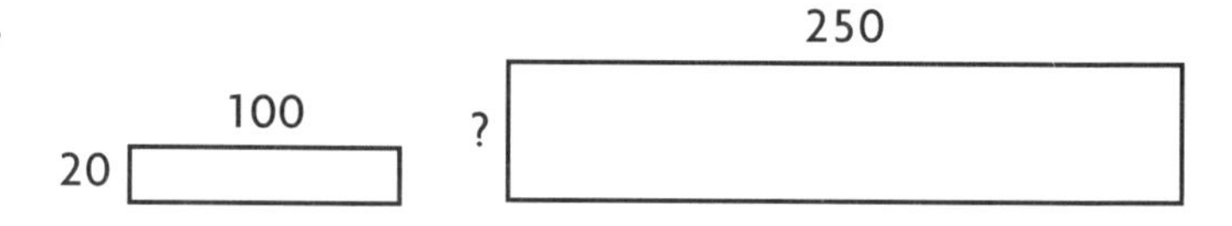

____

6.

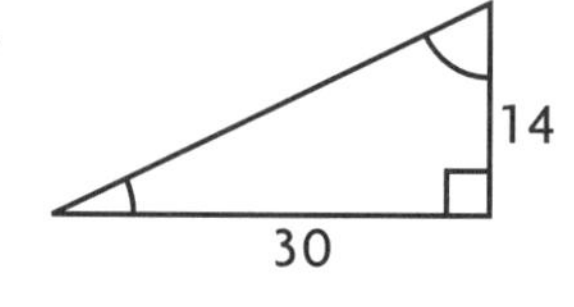

____

7.

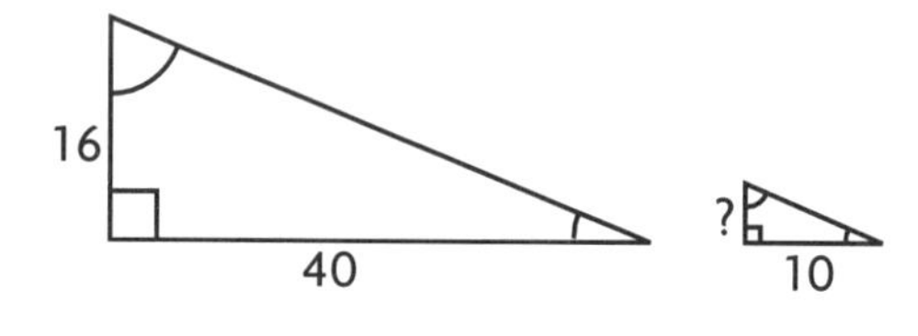

____

Mixed Applications

8. A 10-pound slab of clay will make 15 12-inch flower pots. How many 6-inch flower pots will a slab of clay make?

____

9. Karen makes payments of $199 per month for 3 years for her car. What is the total amount Karen pays for her car?

____

Name ______________________________

LESSON 27.5

Problem-Solving Strategy

Write a Number Sentence

Write a number sentence to solve.

1. Apu had a 3 in. × 5 in. photo that he wanted to enlarge. What is the width of the enlargement?

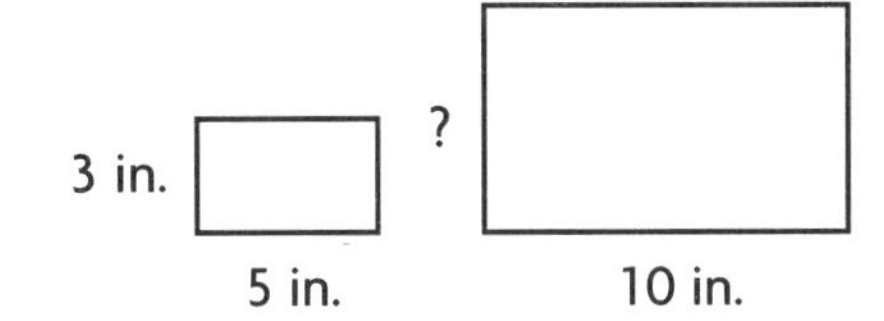

____

2. Joshua built a model plane that was $\frac{1}{25}$ the size of the actual plane. If the model was 2 feet long, how long was the actual plane?

____

3. Jane ran 1 mile south, 2 miles east, 1 mile north, and then returned to her starting point. How far did Jane run?

____

Mixed Applications

Solve.

CHOOSE A STRATEGY

- **Write a Number Sentence**
- **Draw a Diagram**
- **Act It Out**
- **Work Backward**

4. Diana has a dining room table that is 8 feet long and 4 feet wide. She buys a tablecloth that is a similar rectangle and is 12 feet long. How wide is the tablecloth?

____

5. Majdeh bought a coat for $48.99, a scarf for $24.95, and a blouse for $19.95. She received $6.11 change. How much money did she give the cashier?

____

6. Alfonse recorded the noon temperature for five days. On Monday and Thursday the temperature was 94°F. On Tuesday it was 78°F, on Wednesday it was 84°F, and on Friday it was 98°F. What was the mean temperature?

____

7. Yvette is painting her room. She has one wall left that is 9 ft × 14 ft. She has one quart of paint that covers 350 square feet. Will she be able to put two coats of paint on the wall?

____

Name ____

LESSON 28.1

Understanding Percent

Vocabulary

Fill in the blank.

1. ____________________ means "per hundred."

Use counters to show the following on your 10×10 grids. Draw a picture and write the percent.

2. 67 cents out of 1 dollar

____

3. 16 sheep out of 100 animals

____

4. 58 girls out of 100 children

____

Look at the grid. Write the percent that is shaded.

5.

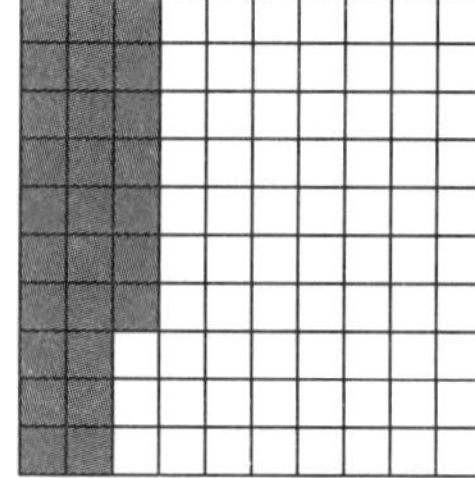

Percent ________

6.

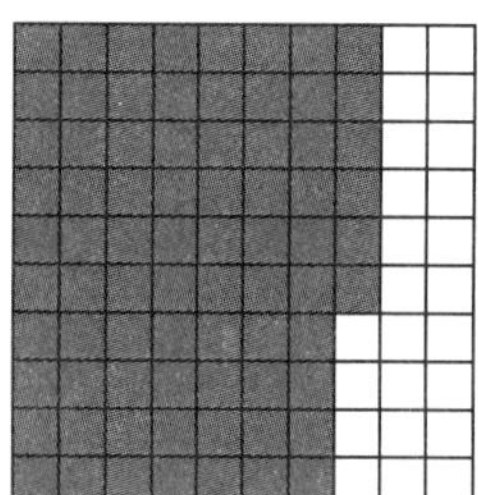

Percent ________

7.

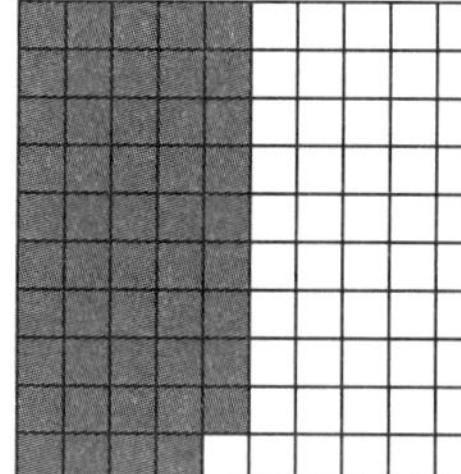

Percent ________

For Exercises 8–9, choose the more reasonable percent. Circle *a* or *b*.

8. "*About half* the students bring their own lunches to school," said the cafeteria worker.

a. 55 percent

b. 85 percent

9. "*Very few* children are sent to the principal's office," said the teacher.

a. 98 percent

b. 2 percent

Mixed Applications

10. Lydia shot 100 free throws and made 84%. How many free throws did Lydia make?

____

11. Brian gave the cashier \$1.00 for his fruit juice. He received \$0.11 in change. What percent of his dollar did he spend on the juice?

____

Name ____________________

Connecting Percents and Decimals

For Exercises 1–4, study the shaded parts of the grid.

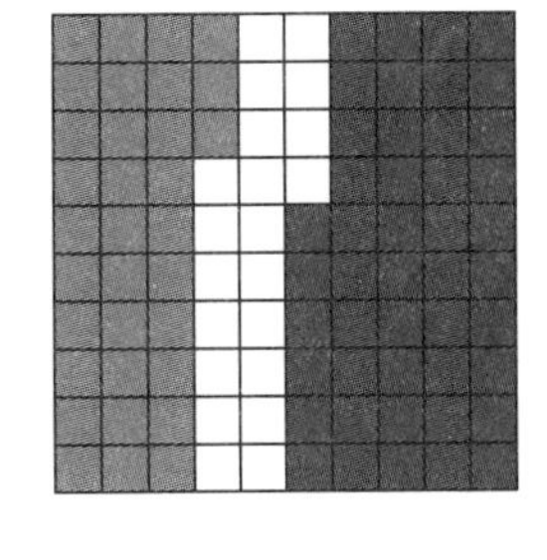

1. What percent of the squares have light shading? __________

2. What percent of the squares have dark shading? __________

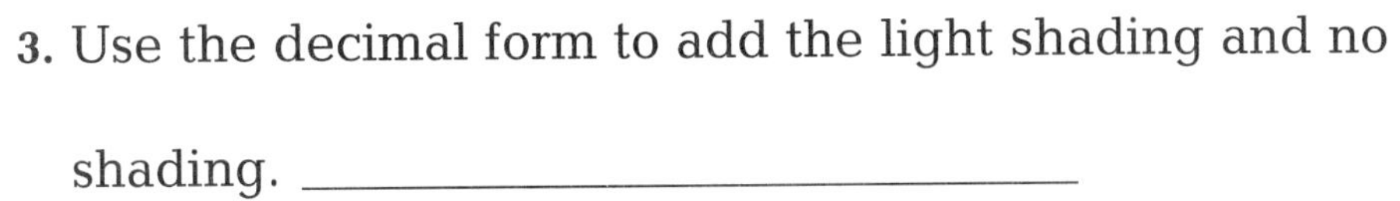

3. Use the decimal form to add the light shading and no shading. __________

4. When you add the light shading, dark shading, and no shading, what percent is the total? __________

Write the number as a percent and as a decimal.

5. seventy-five hundredths __________

6. ninety-three hundredths __________

7. fifteen hundredths __________

8. thirty hundredths __________

Write the decimal as a percent.

9. 0.46 __________

10. 0.79 __________

11. 0.20 __________

12. 0.03 __________

13. 0.18 __________

14. 0.86 __________

Write the percent as a decimal.

15. 38% __________

16. 74% __________

17. 2% __________

18. 16% __________

19. 22% __________

20. 91% __________

Mixed Applications

For Problems 21–22, write the answer as a percent and as a decimal.

21. Sharon and Pete participated in a 100-mile fund-raising relay race. They ran a combined 23 miles. What part of the race did they cover?

____

22. Melanie received $100 for her birthday. She spent $38 on a new dress. What part did she spend on the dress?

____

Name ______________________

LESSON 28.3

Connecting Percents and Fractions

Write the percent as a fraction in simplest form.

1. 50% ______
2. 30% ______
3. 25% ______
4. fifty-six percent ______
5. twenty-three percent ______

Write the fraction as a percent.

6. $\frac{19}{100}$ ______
7. $\frac{8}{25}$ ______
8. $\frac{71}{100}$ ______
9. $\frac{2}{100}$ ______
10. $\frac{2}{10}$ ______
11. $\frac{1}{4}$ ______

Write as a decimal and as a fraction in simplest form.

12. 12 percent ______
13. 24 percent ______
14. 37 percent ______
15. 45 percent ______
16. 60 percent ______
17. 85 percent ______
18. 75 percent ______
19. 50 percent ______

Mixed Applications

20. Mr. Downing went on a 100-day archaeological expedition. He traveled $\frac{1}{5}$ of the days. What percent of the days did he not travel?

______

21. Ken realized that it had been 100 days since school started. He counted 32 days off for holidays and weekends. What percent of days had school been in session?

______

22. Brenda drank one-half gallon of milk a week. Linda drank 8 cups of milk a week. Who drank more milk? Explain.

______

______

23. The ratio of coaches to players is 1 to 12. There are 9 coaches. How many players are there?

______

Name ________________________________

Benchmark Percents

Vocabulary

Fill in the blank.

1. A ______________________ is a commonly used percent that is close to the amount you are estimating.

For Exercises 2–4, choose from the following benchmarks: 10%, 25%, 50%, 75%, and 100%.

2. 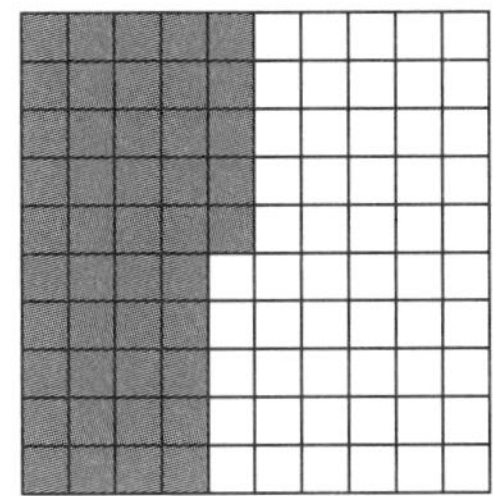

____

3. 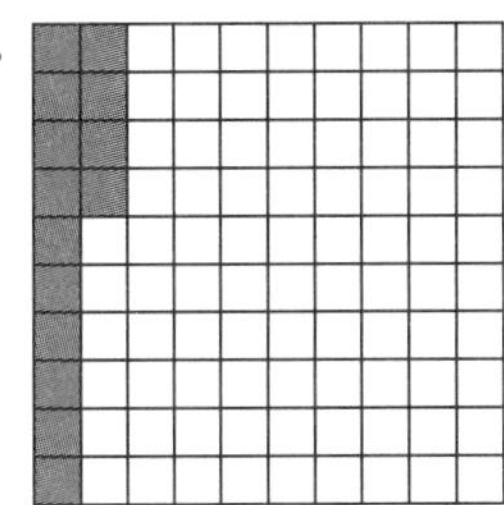

____

4. 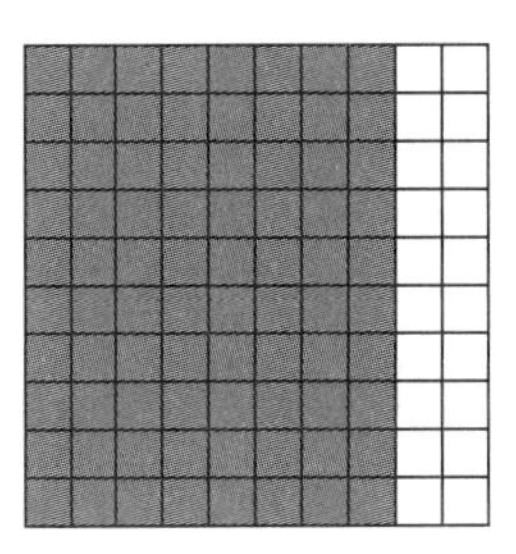

____

Tell what benchmark percent you would use to estimate each percent.

5. 42% __________ **6.** 19% __________ **7.** 78% __________ **8.** 91% __________

For Exercises 9–10, choose the more reasonable benchmark percent. Circle *a* or *b*.

9. *Most* of the class preferred recess to math.

a. 75%

b. 25%

10. *Very few* of the students failed the science exam.

a. 40%

b. 10%

Mixed Applications

11. Ben has finished reading nearly half of his book. What benchmark would you use to estimate the percent of the book he has read?

____

12. Hasan earns $72 for 6 hours of work. How much does he earn for 4 hours of work? for 1 hour?

____

Name ____________________

LESSON 28.5

Percents in Circle Graphs

For Exercises 1–2, use the first circle graph.

STUDENTS WHO HAVE PETS

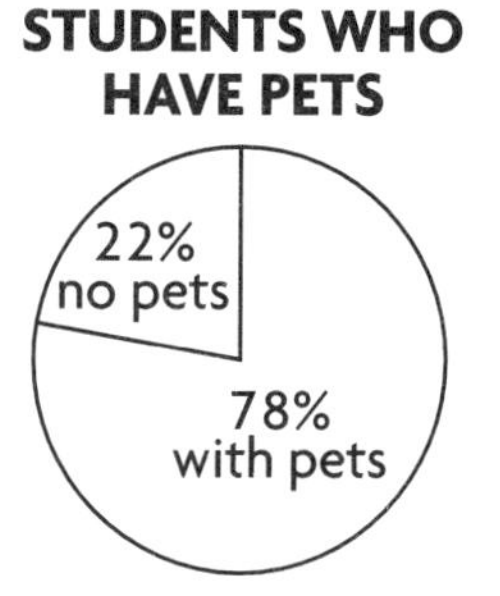

1. Do more than three quarters or less than three quarters of the students surveyed have pets? ____________

2. What decimal represents students who have pets? ________

For Exercises 3–5, use the second circle graph.

TYPES OF PETS

Other 25%
Cats 25%
Birds 22%
Dogs 28%

3. Do more students have dogs or have other types of pets?

____

4. Which groups in the circle are the same size?

____

5. Which two categories cover less than one half of the circle graph?

____

Mixed Applications

For Exercises 6–10, use the table.

FAVORITE DESSERTS	
Ice cream	20 votes
Lemonade	10 votes
Cookies	20 votes

6. What decimal represents the students who chose ice cream? What percent of the students chose ice cream? (HINT: 20 out of 50 chose ice cream.)

____

7. What decimal and percent represent the students who chose lemonade?

____

8. What decimal and percent represent the students who chose cookies?

____

9. If you were to make a circle graph of these data, what would the smallest section show?

____

10. How many times as large would the section for ice cream be as the section for lemonade?

____

Name ________________________________

Problem-Solving Strategy

Make a Graph

Make a graph and solve.

1. Abigail surveyed the fifth-grade students to find out their favorite TV shows. She organized the data in the table at the right. What is the best way for her to display the data? Which TV show is most popular?

____

____

FAVORITE TV SHOWS	
Show	**Percent of Votes**
Plimpton	20%
Queen of the Hill	40%
Atlas	10%
Harborwatch	10%
The Butler	20%

Mixed Applications

Solve.

CHOOSE A STRATEGY

- Make a Model
- Make a Graph
- Guess and Check
- Write a Number Sentence

2. Tamala recorded the average temperature for 6 months. She recorded 48° in April, 59° in May, 69° in June, 76° in July, 74° in August, and 64° in September. Which month had the highest temperature? How can you show this?

____

3. Mylan spent $3 on a magazine. He spent half of his remaining money on a video game. He then spent half of his remaining money on a book. He had $12 left. How much money did Mylan begin with?

____

4. A dog pen will have a length of 18 feet and a width of 12 feet. One length will be formed by the side of a garage. The other three sides will be made of fencing. How much fencing is needed?

____

5. There were 63 people in a hotel. Then 7 checked out, and 3 times that number checked in. How many people are in the hotel now?

____